Commentary

on model form of

General Conditions of Contract MF/1

Home or overseas contracts for the supply of electrical, electronic or mechanical plant
–with erection

A practical guide for users of MF/1

2001 Edition
09/01

Table of contents

1. Introduction

General

This Commentary is not intended to be an authoritative interpretation of the MF/1 Conditions. Nor is it the intention that it should provide interpretation of provisions found in the other model contract documents making up the MF/1 publication. Interpretation is for courts of law or arbitration tribunals. The Commentary is, however, intended to give users practical help and guidance on the clauses of the MF/1 Conditions and the MF/1 contract documents and their interrelationship with each other.

For brevity, Model Form MF/1 and the Commentary on MF/1 are referred to simply as "MF/1" and "the Commentary", respectively, throughout this publication.

MF/1 is suitable for lump sum contracts for the supply of electrical, electronic or mechanical plant with erection. Several sets of additional suggested Special Conditions are also included: those introduced as a consequence of changes in United Kingdom statute not necessarily being required in certain overseas contracts. See section 2 of this Commentary for further comments on UK legislation.

MF/1 initially replaced two model forms in the earlier "A" series model forms (Model Form A and Model Form B3) and subsequently replaced Model Form E as well.

This new edition of the Commentary takes account of the amendments to MF/1 that were included in the 2000 Edition of MF/1. It includes edited content from the two Supplements to the Commentary, *S1-MF/1 (COM)* and *S2-MF/1 (COM)*, which were issued, respectively, in April 1998 and March 1999 and comment previously included in Amendment Slip *MF/1, A/S1* (issued in July 2000).

MF/1 reflects an accommodation of the views of purchasers, engineers and manufacturing industry and provides a fair balance between Contractor and Purchaser. It is hoped that the form will be a suitable model for most modern contracts and that it provides an appropriate measure of flexibility for the parties in the use of "Special Conditions" for particular requirements. The aim has been to provide a basis for uniformity, which will enhance confidence in all users by encouraging fair and equitable treatment, by increasing economy in tendering and by providing a sound basis for competition and quality.

An aide-mémoire to the preparation of the Special Conditions is included in addition to suggested detailed additional Special Conditions which may be considered appropriate where the Works include the provision by the Contractor of computer hardware and software, eg. where the Works include a computerised control or monitoring system. Special Conditions are now also suggested for use where payments are to be determined by measurement (the subject of the 1993 Supplement to MF/1), together with suggested Special Conditions providing (a) for sectional completion and damages for delay in completion of Sections, (b) for contracts which are subject to the Housing Grants, Construction and Regeneration Act 1996 and (c) for contracts which are subject to the Contracts (Rights of Third Parties) Act 1999.

MF/1 also provides a suggested model Form of Sub-Contract, suitable for use where the Main Contract is subject to MF/1 Conditions, and a range of additional model contract documents. The latter include the suggested forms of a Variation Order, a Taking-Over Certificate and a Notice of Delegation of Authority -as well as a Form of Tender, a Form of Agreement, a Form of Defects Liability Demand Guarantee and a Form of Performance Bond.

Features of the MF/1 Conditions

A central philosophy of MF/1 is that the general conditions and any associated Special Conditions negotiated for the particular Contract should alone define the boundaries of the parties' responsibilities towards each other and exclusively provide the parties with their remedies against each other. The Contract alone (and not some extraneous rules of law) should govern the parties' relationship. It is important that the parties should recognise that there is no scope for implying remedies or rights or responsibilities into the Contract. Any such rights, remedies or responsibilities must be specifically included if they are to be relied upon. This philosophy is extended to the suggested MF/1 Form of Sub-Contract.

The MF/1 Conditions specifically require the Engineer to act fairly between the parties and require any restrictions on his powers and duties under the Contract to be disclosed at the tender stage so that, in particular, the Contractor is put on notice as to the extent to which the Engineer may, without the Purchaser's consent, order variations, grant extensions of time or involve the Purchaser in extra Cost. Strict time limits are introduced for the claiming of extensions of time and prompt notification is required of all claims by the Contractor for extra Cost. The provisions for liability for accidents and damage clearly apportion responsibility between Contractor and Purchaser and the provisions for insurance of the Works more accurately reflect the risks for which insurance is available under "Contractors All Risks" insurance policies.

MF/1 contains no detailed terms of payment, these being left for negotiation between the parties although, in the aide-mémoire to the preparation of the Special Conditions, suggested clauses are given for progress certificates of payment and terms of payment.

Greater emphasis has been placed on the Contractor's responsibility for design and provisions included to deal with cases where the Purchaser, Engineer and Contractor are all involved to some extent in the design of the Works.

The MF/1 Conditions specifically require the Contractor to produce a Programme. Completion is related to the passing of the Tests on Completion and the issue of a Taking-Over Certificate, and provision is included for Performance Tests which are to take place after taking-over. The parties can agree their own Defects Liability Period for the Works and only if no other is agreed is the period 12 months. The Contractor is required to accept some liability for latent defects.

Unless arbitration proceedings have been commenced within three months after the issue of the final certificate of payment, that certificate will, except in certain circumstances, take effect as a discharge to the parties of their obligations under the Contract.

Parties to contracts under MF/1 Conditions will require access to the information contained in other bodies' publications which is included by reference in the Conditions, particularly in the field of dispute resolution, and sources for such publications are suggested in section 8 of the Commentary.

A detailed list of the changes to the 1995 reprint of MF/1, MF/1 (rev 3), is given in Amendments List *MF/1, A/L3* which is available from the IEE at Savoy Place or from the IEE Website:-

http://www.iee.org.uk

2.   Legal matters

As with the previous model forms which it replaces, MF/1 has been drafted against a background of UK legal systems. Thus it was thought helpful to refer to certain UK statutes –particularly those aimed directly at the engineering construction industry. Table 1 below gives the locations in MF/1 or the Commentary of those references in the belief that that will prove useful to overseas users.

MF/1 is suitable for both home and overseas contracts. "Home contracts" include those contracts governed by English & Welsh law (referred to widely and in MF/1 and this Commentary simply as "English law"), Scots law or Northern Ireland law and users should be mindful that not all UK statutes are intended to have effect in every home jurisdiction. For example, the Arbitration Act 1996 does not extend to Scotland. Thus, although drafted on the basis of English law, MF/1 may, with care, be simply adapted for use (for example, under Scots law) and it may also be adapted, but only with appropriate legal advice, for use under other applicable laws.

Table 1: UK statutes referred to in MF/1 or in the Commentary
Note: The following abbreviations have been used in the table.

(i)	Co	— Commentary
(ii)	FoA	— MF/1 Form of Agreement
(iii)	FoT	— MF/1 Form of Tender
(iv)	gc	— MF/1 general conditions
(v)	SC	— MF/1 Special Conditions
(vi)	S-C	— MF/1 Form of Sub-Contract
(vii)	sc	— sub-clause

The Unfair Contract Terms Act, 1977	(Co, sc 36.9.)
The Highways Act,1959	(Co, sc 21.1 - 21.4.)
The Construction (Design and Management) Regulations, 1994	(Co, section 2.)
The Arbitration Act,1996	(gc & Co, sc 52.5; S-C, sc 19.4.)
The Housing Grants, Construction and Regeneration Act, 1996, Part II*	(SC & Co, section 5.)
The Late Payment of Commercial Debts (Interest) Act, 1998	(Co, sc 40.2)
The Contracts (Rights of Third Parties) Act,1999	(Co, section 5; SC)
The Administration of Justice (Scotland) Act, 1972	(SC, sc 54.1.)
The Law reform Miscellaneous Provisions (Scotland) Act, 1990	(SC, sc 52.5.)

*and its related statutory instruments (ie. "Schemes") and Northern Ireland Order.

Note: All references to "Value Added Tax" are in consequence of the V.A.T. Act, 1972 (FoT; FoA.)

A note on the UK Construction (Design and Management) Regulations 1994
It was not found necessary to include a specific provision in the MF/1 Conditions as a consequence of the introduction in the UK of the above Regulations ("the CDM Regulations") and this is explained below.

In most cases the construction of the Works in the United Kingdom will be subject to the CDM Regulations. The CDM Regulations require the Purchaser to appoint a "planning supervisor" and a "principal contractor". The planning supervisor is responsible for ensuring that all those responsible for the design of any part of the Works allow in their design for health and safety risks both in construction and operation. The principal contractor is responsible for ensuring the development of the "health and safety plan" for the construction work on Site.

Changes requested by the planning supervisor or by the principal contractor for the purposes of the CDM Regulations could involve the Contractor or his Sub-Contractor(s) in extra Cost. MF/1 will not allow the possibility of recovery of such extra Cost unless arising from a variation order (which can only be issued by the Engineer) or an Engineer's instruction or decision under the Contract In essence, MF/1 provides for the Engineer to be the sole channel of communication between Purchaser and Contractor for the purposes of the Contract.

There is thus no role within the administration of the Contract for the planning supervisor or the principal contractor of the CDM Regulations. This in no way means that the parties should not observe the requirements of the CDM Regulations - they must, of course, do so and the Contractor will be deemed to have allowed in the Tender for the cost of complying with the regulations. Nevertheless, if compliance with the CDM Regulations necessitates a variation of the Works or the giving of instructions to the Contractor these must come from the Engineer and no one else.

3.   General conditions

1.   DEFINITIONS AND INTERPRETATIONS
This clause consists of a series of key definitions used throughout the Conditions and which should also be used throughout the contract documents, not least the Specification. Many of the definitions are self-explanatory and call for no comment.

In accordance with the convention adopted, all principal words of defined terms have capitalised initial letters wherever they appear in the text (except where, under sub-clause 1.5 (Headings and marginal notes), this is made unnecessary) so that the user is reminded that such terms have no other meaning than that defined in clause 1. This convention is also used throughout the Commentary. It should be noted that differing sets of defined key terms are used in the Contract and the Sub-Contract though some key terms are common to both of them.

1.1.a   The Purchaser is identified in the Special Conditions and effectively his identity cannot change except with the Contractor's consent which, by sub-clause 1.4 (Notices and consents), may not be unreasonably withheld (see comments below). The Contractor's consent is required because it is unreasonable to expect him to accept a new "paymaster" without an opportunity to make appropriate enquiries.

1.1.b   No provision is included for the Purchaser to consent to an assignee of the Contractor since, in general, the Contractor will have been selected for his ability to perform the Contract in question and, in consequence, the Purchaser should not be required to accept performance of the Contract by some third party.

1.1.c   The definition encompasses not only Sub-Contractors who are named in the Contract for the execution of particular parts of the Works but also those to whom the Contractor may have sub-let parts of the Works with the Engineer's consent. Again, it should be noted that the Engineer's consent must not be unreasonably withheld - see also the comments on sub-clause 3.2 (Sub-contracting).

1.1.d   The name of the Engineer must be specified in the Special Conditions. The Engineer may be a corporate body, a partnership or an individual. If no Engineer is specified, then the Purchaser must fulfil that role himself. The latter is not necessarily a desirable course of action since the role of the Engineer under the Conditions is the traditional role which requires the Engineer to exercise discretion in relation to a number of matters and to act in an impartial manner, a role which it may be difficult for the Purchaser himself to perform. Where the Engineer is in fact an employee of the Purchaser, it may be necessary to provide in the Special Conditions some further amplification of his role under the Contract - see comments on clause 2 (Engineer and engineer's representative) below.

1.1.g   The definition of "Contract" establishes and identifies the documents which form the agreement between the parties for the execution of the Works and to which reference will have to be made in construing the rights and obligations of the parties. It is most important that any alterations to the Conditions, the Specification, drawings, schedules and Tender which are agreed in the negotiations are expressly incorporated into or referred to in the Letter of Acceptance since the rules of English law relating to the construction and interpretation of the Contract make it very difficult for the parties to establish as terms of the Contract matters agreed in the negotiations which are not incorporated expressly or by reference.

1.1.h   The Contract Price is of course not necessarily the total sum the Contractor will receive for the execution of the Works since the Conditions provide for adjustment to the Contract Price by way of addition or deduction in a number of cases, eg. for variations and claims for extra Cost. The "Contract Price" is thus a reference point for establishing what is ultimately due to the Contractor.

1.1.i   The Contract Value is an expression used to enable a value at any particular time to be assigned to parts of the Works. It includes, therefore, not only a proportion of the Contract Price but also additions or deductions other than for escalation. It is important, for example, in determining the amounts to be included in an interim certificate of payment where the Contractor becomes entitled to payment for parts of the Works the delivery or erection of which has been suspended under clause 25 (Suspension of work, delivery or erection).

1.1.l   "Letter of Acceptance" is used to describe the Purchaser's formal acceptance of the Contractor's proposal for the execution of the Works. It should constitute and take effect as a binding acceptance. It should therefore contain a complete list of the documents which comprise the Tender together with confirmation of any alterations thereto and any alterations to the general conditions or Special Conditions, Specification and drawings. It should also, of course, state the Contract Price. The Letter of Acceptance may be preceded by a letter of intent by which the Purchaser expresses his intention to award the Contract to the Contractor. Letters of intent take many forms and do not usually constitute a binding acceptance of the Contractor's Tender. If a letter of intent has been issued, it, too, should be referred to in the Letter of Acceptance.

1.1.m   The Time for Completion runs from whichever is the later of the three dates specified. For the provisions of sub-paragraph (c) to be effective, it is essential that any legal, financial or administrative requirements are clearly defined in the Special Conditions or elsewhere in the Contract. The Time for Completion ends on taking-over subject to any extension of time which the Contractor may have been granted under sub-clause 33.1 (Extension of time for completion). If the Contractor fails to complete within the Time for Completion, the Purchaser may be entitled to damages for delay under clause 34 (Delay).

1.1.t   Alterations to the general conditions to be set out in the Special Conditions include any additional clauses and the deletion of or amendment to clauses in the general conditions. The intention is that all changes to the general conditions should be embodied in the Special Conditions; the general conditions themselves remaining unchanged.

1.1.u   The Site describes only the areas which are made available by the Purchaser to the Contractor for the purposes of the Works. It may conveniently be defined in the contract documents by reference to an appropriate plan.

1.1.v   The Tests on Completion are those tests which when successfully concluded will lead to the issue of the Taking-Over Certificate and the acceptance of the Works by the Purchaser. Under the Conditions the Tests on Completion do not include "Performance Tests" which are dealt with separately. The definition assumes that full details of the Tests on Completion will be agreed at the tender stage and that they will be described in detail in the Contract so that there can be no argument as to the tests which the Works are expected to pass before the Works are taken over. However, in some cases, particularly where the Contract may involve the supply of computer software and hardware, it may not be possible to agree these in detail at the tender stage and such tests cannot therefore be specified in the Contract. Accordingly the definition permits later agreement of the Tests on Completion and, if this be the case, a detailed specification of the tests should be incorporated into the Contract by way of an appropriate variation order.

1.1.w   The Conditions envisage that the Performance Tests will be carried out after taking-over. The Performance Tests should be detailed in the Specification or in an appropriate performance tests schedule which if agreed after the date of Contract should be incorporated into the Contract by a variation order. See the detailed comments on clause 35 (Performance tests).

1.1.x   Provision is made in the Conditions for the parties to agree a Defects Liability Period by completing the appropriate provision in the Special Conditions. In the absence of agreement the Defects Liability Period will be the customary period of 12 months after taking-over.

1.4   Notices and consents
This sub-clause makes it clear that in order to be considered a proper communication for the purposes of the Contract, any notice or consent must be in writing. It further makes it clear that where under the Conditions the consent of any of the parties or of the Engineer is required, such consent is not to be unreasonably withheld.

1.5   Headings and marginal notes
The purpose of this sub-clause is to make it clear that the headings and marginal notes to the clauses are for ease of reference only and may not be used to assist in the interpretation or construction of the Contract.

2.   ENGINEER AND ENGINEER'S REPRESENTATIVE
This clause defines the Engineer's role for the purposes of the Contract. The clause does not define or determine the relationship between the Engineer and the Purchaser. Where the Engineer is an independent consultant, his authority to commit the Purchaser to extra Cost, and his duties will be defined in the contract between the Engineer and the Purchaser relating to the Engineer's appointment.

2.1 Engineer's duties

The Engineer is not a party to the Contract but his role is to act on behalf of the Purchaser as the essential channel of communication between the Purchaser and the Contractor and to perform the duties that are assigned to him under the Contract. Generally the Engineer is authorised by the Purchaser to do all things within his power to ensure satisfactory completion of the Works in accordance with the Contract. His duties will therefore include the issue of instructions and information so as to enable the Contractor to design the Plant, review and comment on the Contractor's detailed design of the Works and generally the Engineer must watch, supervise and test the Works and the Plant so as to ensure that materials and workmanship are as specified. The Engineer must also satisfy himself that any claims for payment or for additional Cost made by the Contractor are justified under the terms of the Contract.

Although under the Contract the role of the Engineer is to perform these duties, it is likely, particularly where the person appointed to act as Engineer is not an independent Engineer but an employee of the Purchaser, that the Purchaser will have imposed restrictions on the Engineer in the performance of his duties. For example, the Engineer may not be authorised by the Purchaser to issue instructions or orders which may result in the Contractor claiming extra Cost, nor, perhaps, to certify the Contractor's entitlement to an extension of time, without first obtaining the Purchaser's specific approval. Such restrictions on the authority of the Engineer are becoming more common and are sometimes imposed on independent consulting engineers. The sub-clause provides that if there are any restrictions of this nature imposed on the Engineer under the terms of his appointment full particulars must be set out in the Special Conditions. The purpose of this is to ensure that the Contractor can take these restrictions into account when framing his Tender.

2.2 Engineer's representative

This sub-clause allows the Engineer to appoint one or more representatives. Normally such representatives will be employed upon the Site as "resident engineers". Unlike civil engineering contracts, much of the Contractor's work will be carried out in his own or his Sub-Contractors' premises. Since the Contract may provide for detailed and specific tests of Plant when in the Contractor's works or immediately before delivery, it is normal practice for the Engineer to appoint one or more of his own staff as his representative to carry out any necessary supervision of the Plant in the course of manufacture and its testing and examination at the Contractor's premises. When the Contractor's or the Sub-Contractor's factory is overseas it is not unusual for the Engineer to appoint as his representative an independent firm within relatively easy reach of the premises, who will be responsible for supervision and inspection, examination and testing of the relevant Plant whilst in the course of manufacture. The Engineer should give details of any such appointment and of the representative's delegated duties to the Contractor under sub-clause 2.3 (Engineer's power to delegate). The sub-clause sets out the normal responsibilities of an Engineer's Representative, namely to watch and supervise the Works and to test and examine any Plant or workmanship employed. If further authority is to be delegated to him, this must be done under the provisions of sub-clause 2.3 - see comments below.

2.3 Engineer's power to delegate

The Engineer can delegate any of his duties to the Engineer's Representative at any time and he may likewise revoke any such delegation. Any delegation or revocation must be in writing and is not to take effect until the Contractor has a copy. A partially completed example notice of such delegation using the Form of Notice of Delegation of Authority has been provided with MF/1.

Instructions, etc. given by the Engineer's Representative in relation to matters which have been delegated to him have the same effect as if they had been given by the Engineer. If the Contractor disagrees with any decision, instruction or order that he has received from the Engineer's Representative, he can refer the matter to the Engineer himself who is empowered to reverse or vary the decision, etc. in accordance with sub-clause 2.6 (Disputing engineer's decisions, instructions and orders) - see below.

2.4 Engineer's decisions, instructions and orders

This sub-clause requires the Contractor to proceed in accordance with the decisions, instructions or orders that he receives from the Engineer provided that such decisions, etc. are made in accordance with the Conditions. Subject to his rights under sub-clause 2.5 (Confirmation in writing) to require the Engineer to confirm in writing any oral instruction, and under sub-clause 2.6 (Disputing engineer's decisions, instructions and orders) to dispute any written instructions, the Contractor must proceed in accordance with such decisions, etc.

2.5 Confirmation in writing
If the Contractor receives an oral decision, instruction or order he is entitled to require the Engineer to confirm it in writing before proceeding, provided that he makes his request without any undue delay. The Contractor is entitled to refrain from carrying out the decision, instruction or order until he has received such written confirmation.

2.6 Disputing engineer's decisions, instructions and orders
The Contractor can dispute any written decision, instruction or order of the Engineer provided he does so by written notice within 21 days after receiving it. The notice must give the Contractor's reasons for disputing it. The Engineer then has a further period of 21 days within which to confirm, reverse or vary his decision, instruction or order and he too must give reasons for his response. It should be remembered that, notwithstanding that the Contractor may have disputed the Engineer's decision, etc. he is still required by the Contract to proceed with the Works in accordance with the decision, etc. in dispute.

Once the Engineer has given his response, then, if either party wishes to dispute the response, they have a further period of 21 days within which to refer the matter to arbitration. If they do not do so within that period, then the decision, etc. of the Engineer as confirmed, reversed or varied in accordance with this sub-clause 2.6 will be final and binding on the parties unless the Contract is subject to the Housing Grants, Construction and Regeneration Act 1996 (the Act) and the decision, etc. is overturned by an adjudicator appointed in accordance with the Act. See section 5 of this Commentary for discussion of the application of the Act.

2.7 Engineer to act fairly
In all cases where the Engineer is required to exercise his discretion under the Contract, the Engineer is under a duty to act and to exercise that discretion fairly. In this area the Engineer is, under the Contract, not simply an agent or servant of the Purchaser. The requirement of fairness and impartiality is an essential feature of his role.

2.8 Replacement of engineer
The Contractor will have taken the identity of the Engineer into account when making his offer to the Purchaser to construct the Works because he relies upon the ability and professional judgement of the Engineer to act fairly and impartially in carrying out his duties under the Contract. In the course of the Contract, however, circumstances may arise where it is necessary for the Purchaser to replace the Engineer. The Engineer is named in the Contract and as such his replacement requires the consent of the Contractor because the Purchaser cannot unilaterally alter a term of the Contract. By reason of the provisions of sub-clause 1.4 (Notices and consents), the Contractor may not unreasonably withhold that consent.

3. ASSIGNMENT AND SUB-CONTRACTING

3.1 Assignment
The Contractor is not entitled to assign performance of the Contract to some other contractor since the Contractor will generally have been chosen by the Purchaser because of his ability to perform the Contract. This does not prevent a Contractor from assigning to his bankers his right to payment in connection with any facilities that they may provide to him. Neither does it affect insurers' rights of subrogation where such rights apply. ("Subrogation" is a mechanism whereby an insurer may gain the advantages of the insured's rights and/or benefits by taking over and treating any such rights of the insured, or any such benefits due to the insured, as rights of, or benefits due to, the insurer). In this latter respect it should be noted that when using the MF/1 Form of Performance Bond the Guarantor would not have a right to perform the obligations of the Contractor. This is because it is considered that the Purchaser will have selected a Contractor on the basis of various criteria that a Guarantor would not necessarily be in a position to fulfil.

3.2 Sub-contracting
For similar reasons to the prohibition against assignment, if the Contractor wishes to sub-contract part of his obligations under the Contract then he must first obtain the Engineer's consent. Normally

the Contractor will have been required to indicate in his Tender those parts of the Works that he would wish to sub-contract and the Purchaser may indeed have required him to place a particular sub-contract with someone of the Purchaser's own choice. If these Sub-Contractors are named in the Contract, then consent is not required. The Contractor does not have to obtain consent for the purchase of materials or for minor details.

In all cases the Contractor will be responsible for work performed by Sub-Contractors as if he had done the work himself even if the Sub-Contractor concerned has been chosen or nominated by the Purchaser. A Contractor who has any doubts about the ability of a Sub-Contractor whom he is required by the Purchaser to use should therefore make his position abundantly clear at the tender stage and possibly seek to limit his otherwise total responsibility for the work in question.

4.   PRECEDENCE OF DOCUMENTS

4.1   Precedence of documents

Since the whole of the terms of the Contract cannot necessarily be ascertained from studying a single document but rather from a number of documents, including the Conditions, the Specification and drawings, it is important to establish an order of priority. More than one document may seek to deal with the same matter and discrepancies may have arisen in consequence of the negotiations. It is important, therefore, to know when interpreting the Contract the order of priority of documents. If no order is specified in the Contract, the general rule of English law will apply, namely that the specific will take precedence over the general. Thus, a clause in the Special Conditions agreed in the negotiations will, if it conflicts with the wording of a clause in the general conditions, take precedence.

5.   BASIS OF TENDER AND CONTRACT PRICE

5.1, 5.2 and 5.3   Contractor to inform himself fully and Site data

The purpose of these sub-clauses is to make it clear that the Contractor has prepared his Tender for the execution of the Works on the basis of his own enquiries as to matters affecting the Site, if access has been made available to him, and in accordance with his own interpretation of the invitation to tender, the Conditions, the Specification, drawings, the schedules, plans and on information made available to him in writing by the Engineer or the Purchaser. The Contractor is also required so far as he can to have satisfied himself as to the safety regulations that will be applicable to Works in the country where the Plant is to be erected. If any written information made available to the Contractor by the Purchaser or Engineer is incorrect, the Contractor may be able to recover the Cost of any consequential alterations to the Works under the provisions of sub-clause 16.2 (Errors in drawings, etc. supplied by purchaser or engineer) - see below.

5.4, 5.5 and 5.6   Provisional sums and Prime cost items

The expression "provisional sum" is used in the Conditions to describe a sum which has been included in the Contract Price to cover contingencies or the purchase of items or services, the need for which has been anticipated by the Purchaser or the Engineer but the exact quantities or amounts of which could not be determined at that time.

The expression "prime cost items" is used in the Conditions to describe bought-in items of Plant or services which have been obtained from a supplier (who may be named in the Contract, or nominated by or otherwise approved by the Engineer) for incorporation into the Works by the Contractor. Payment is made at the net cost to the Contractor, the "prime cost" (normally supported by the supplier's invoice) plus the agreed Contractor's margin which must be nominated in the Appendix to the Contract. It is normal to define in the Contract the Plant and services which will be paid for on this basis.

Work covered by a provisional sum may or may not be the subject of a quotation from the Contractor or from a Sub-Contractor. A provisional sum may only be used in accordance with the Engineer's instructions. If a provisional sum is not used at all, then, in determining the amount to which the Contractor is entitled under the final certificate of payment, it will be deducted in its entirety. Prime cost items are treated in a similar manner, namely, that the Contractor is entitled to be paid only the actual cost plus any percentages for profit and attendance stated in the Appendix.

Sub-clause 5.6 makes it clear that if prime cost items are used to enable work to be done or Plant to be supplied otherwise than by the Contractor, the Contractor is to have no responsibility for such Plant or work done unless he has approved in writing the person to supply it.

5.7 Unexpected site conditions

Whilst the Contractor is deemed under sub-clause 5.1 (Contractor to inform himself fully) to have inspected the Site if access has been made available to him, this sub-clause permits him to recover the extra Cost of dealing with unexpected Site conditions, the nature of which could not reasonably have been ascertained by the Contractor by his inspection of the Site or from the information that has been made available to him for the purposes of the Tender. Under sub-clause 41.2 (Allowance for profit on claims) the Contractor is entitled in addition to be paid a profit element. The Contractor must inform the Engineer of any steps he proposes to take to deal with the situation and obtain the Engineer's approval thereof. A provisional sum may be included in the Contract Price as a contingency against the Cost of dealing with unexpected Site conditions.

6. CHANGES IN COSTS

6.1 Statutory and other regulations

Changes in legislation can affect the Cost to the Contractor of performing his obligations under the Contract either upwards or downwards. This sub-clause permits the Contract Price to be adjusted accordingly, provided the relevant changes in legislation were made after the date of the Tender.

6.2 Labour, materials and transport

It is intended that where the Contract Price is to be adjusted according to the rise or fall in the Cost of labour, materials or transport any such adjustment should be calculated in accordance with an appropriate contract price adjustment formula to be stated in the Special Conditions. It should be noted that the sub-clause is to apply unless it is specifically excluded by the Special Conditions. In referring to any particular formula, particular care should be taken to ensure that the formula is appropriate to the work in question and identifies the indices which are to be used for the purpose of calculating any adjustments. It may be appropriate to apply different formulae to different Sections or parts of the Works.

7. AGREEMENT

7.1 Agreement

Although under English law the Letter of Acceptance will cause a binding contract to be formed for the execution of the Works it is common practice for one or other of the parties to require a formal agreement to be entered into. If a formal agreement is required, the party who requests it must pay for the costs of its preparation, completion and any necessary stamping. It should be remembered that if the agreement is executed as a deed, this has the effect under English law of extending the limitation period for breach of contract from six years to twelve years. The Form of Agreement referred to in this clause will be found in MF/1.

8. PERFORMANCE BOND OR GUARANTEE

It should be noted that this clause relates only to the MF/1 Form of Performance Bond (or any other contract performance bond that may be specified in the Special Conditions) for use in guaranteeing the performance of the Contract and **not** to the MF/1 Form of Defects Liability Demand Guarantee (or any alternative retention bond).

8.1 Provision of bond or guarantee

The Purchaser may require the Contractor to provide a performance bond for the due performance of the Contract and a suggested Form of Performance Bond can be found in MF/1. Essentially that form of bond is payable only on proof of default evidenced either by judgement of the court or an arbitrator's award or alternatively by a joint submission from the Purchaser and Contractor. Whether or not the form of bond included with the Conditions is used, it is most important that the particulars required by the third paragraph of the sub-clause are included in the Special Conditions.

See also the comment on the Form of Performance Bond which is made in section 7 of this Commentary.

8.2   Failure to provide bond or guarantee

This sub-clause allows the Purchaser to terminate the Contract if the Contractor has not provided the required bond or guarantee within thirty days after the date of the Letter of Acceptance or within such further time as he has been allowed by the Purchaser. If the Purchaser terminates under this provision, then he is only entitled to recover from the Contractor the Cost properly incurred by him incidental to the obtaining of new tenders. He is not entitled to recover the difference between the Contract Price under this Contract and the price bid by the successful contractor when new tenders are called for.

9.   DETAILS CONFIDENTIAL

9.1   Details confidential

This clause is of a general nature and should give neither party any particular difficulty. Both in the negotiations and during the course of performance of the Contract it may be necessary for either of the parties to disclose highly confidential information to the other which they would not otherwise wish to be disclosed.

10.   NOTICES

10.1, 10.2 and 10.3 Notices to purchaser and engineer, Notices to contractor and Service of notices

These three sub-clauses deal with the procedures for service of notice by the parties on each other or to or from the Engineer. In general, notices must be sent to the Purchaser or to the Engineer at the address nominated in the Special Conditions and, in the case of a notice to the Contractor, to his principal place of business or to such address as the Contractor may have nominated for the purpose. Notices may be sent by all modern methods of communication (with the exception of electronic mail) and are deemed to have been served in the case of telex, cable or facsimile transmission at the time of transmission. A notice sent by post will be deemed to have been served four days after posting.

It is for consideration by the parties to the Contract whether notices served by telex, cable or facsimile should be confirmed by letter, or if the safe receipt of such notices should be confirmed by other means.

It is not the intention of this clause to exclude other methods of communication in respect of communications other than notices, for example, direct computer links, which may be agreed between the parties to the Contract. In the case of such agreement, the parties will obviously provide for such means of safeguard and verification as they may deem appropriate.

11 – 12. PURCHASER'S GENERAL OBLIGATIONS

11.1   Access to site

Unless a particular date upon which the Contractor may have access to Site is stated in the Contract, the Purchaser is required to give access in reasonable time. In electrical and mechanical works contracts it would be unusual for the Contractor to need access to the Site from the date of the Contract since it would be normal for much design, manufacture and prefabrication to be carried out otherwise than on Site. The date at which access is required should be stated in the Tender and in the Programme wherever possible. The granting of access requires the Purchaser not only to hand over possession of the Site to the Contractor but also provision of the means of access for all Plant and Contractor's Equipment. There should be no problem at all where road or rail access is already available but in the case of a "green field" site it would be important to state in the Specification any particular requirement relating to access such as load bearing capacity. If access is by rail, any limitations on the size or weight of load should also be specified.

11.2   Wayleaves, consents, etc.

In addition to any consents that may have been obtained from the relevant authorities to construct the Works this sub-clause requires the Purchaser to ensure that all other necessary consents, wayleaves or approvals have been obtained (eg. from local landowners over whose land a right of access may be required or whose consent may be needed to the erection of overhead lines or cables to service the Works). The time by which any consents, wayleaves or approvals are required should have been stated in the Tender or Programme. Failure by the Purchaser to give access in time could involve delay in completion and claims by the Contractor not only for extensions of time but also extra Cost.

11.3 Import permits, licences and duties
It is the Purchaser's responsibility to obtain any import permits or licences in time to enable the Plant to be delivered to Site without any delay on importation. The Programme should specify a time by which such permits or licences should be provided. The Purchaser is also required to pay all customs and import duties on importation. If, in the first instance, the Contractor is required to pay such duties on importation, then the Purchaser must reimburse them and the Special Conditions should so provide.

11.4 Foundations, etc.
The Conditions assume that the Contractor will not be required to carry out any civil engineering work for buildings, structures or foundations, this being the normal practice in electrical and mechanical works contracts. The Contractor will usually be expected to provide the Engineer and the Purchaser with data, for design of any structure required or of any necessary alterations to existing buildings and of foundations for the Plant. If the Contractor supplies design data in good time, delays by the Purchaser may entitle the Contractor to an extension of time and to claim for extra Cost. The times by which the Purchaser is required to complete any necessary civil engineering work should at the very least be stated in the Programme.

11.5 Purchaser's lifting equipment
The Purchaser does not have to make available any lifting equipment that may be available on Site unless the Special Conditions so provide. Where lifting equipment is available, the Purchaser will be required to operate such equipment when so requested by the Contractor but at the Contractor's expense. The Purchaser would retain control of the equipment concerned and be responsible for its safe working.

11.6 Utilities and power
For work on Site the Purchaser is required to make available to the Contractor supplies of electricity, water, gas, air and other services as specified in the Special Conditions. The source of such services should be specified in the Special Conditions and any limitations as to their use.

11.7 Power, etc. for tests on site
If the Contract provides for tests on the Site then the Purchaser is required to provide the necessary power, skilled and unskilled labour, materials, etc., that the Contractor may reasonably require to enable the tests to be carried out effectively. For the avoidance of doubt as to the extent of the Purchaser's obligations the Contractor's requirements should be specified in the Tender.

11.8 Breach of purchaser's general obligations
This sub-clause makes it clear that if the Contractor incurs extra Cost in consequence of the Purchaser being in breach of any of his general obligations under clause 11, the Contractor is entitled to recover the extra Cost and, by reason of sub-clause 41.2 (Allowance for profit on claims), to an allowance for profit in addition.

12.1 Assistance with laws and regulations
Where the Works are to be executed outside the Contractor's country, the Purchaser is required by this clause to give assistance to the Contractor in ascertaining laws, etc. which may affect the Contractor in the performance of his obligations under the Contract. It should be noted that the Purchaser is only obliged to provide assistance, he is not required to identify laws that might be applicable or to give advice thereon.

13 – 22. CONTRACTOR'S OBLIGATIONS
These clauses detail the Contractor's obligations in relation to the Works both whilst the Plant is under construction at the factory and on Site. If more detailed provisions are required, then these should be included in the Special Conditions or in the Specification. Sub-clauses 15.7 (Purchaser's use of drawings, etc. supplied by contractor), 15.9 (Manufacturing drawings, etc.) and 16.2 (Errors in drawings, etc. supplied by purchaser or engineer) have been included here for user convenience, ie.

12

with the other provisions relating to drawings, etc., even though they are not Contractor's obligations.

13.1 Contractor's general obligations
This sub-clause sets out the Contractor's basic obligations under the Contract in relation to the design, manufacture, delivery, erection and testing of the Plant and Works (and, where required, setting to work or commissioning) within the Time for Completion.

13.2 Manner of execution
This sub-clause is of a general nature governing the way in which the Contractor is to execute the Works. If the manner is not described in the Specification, then he must execute the Works to the reasonable satisfaction of the Engineer. All work on Site is to be carried out in accordance with the Engineer's reasonable directions. Any other work to be performed such as testing or quality assurance procedures is also to be carried out in accordance with the Engineer's reasonable directions. The detailed requirements of work of this nature, eg. provisions relating to certification and quality assessment should be included in the Specification and in the Special Conditions.

13.3 Contractor's design
The Contractor's obligation is to provide the detailed design of the Plant and Works in accordance with the requirements of the Specification. If the Contractor's design fulfils those requirements, then he has complied with his obligations under the Contract. It is clear from this that the Purchaser and Engineer must take the greatest possible care in framing their requirements for the Works in the Specification. On his part, the Contractor must accept that he will take responsibility for any design provided by the Engineer or the Purchaser unless he has within a reasonable time specifically disclaimed his responsibility for such designs in writing. This accords with the standard practice of the electrical and mechanical engineering plant industry.

The last paragraph of the sub-clause has been inserted to emphasise that the Contractor's responsibilities for the Works do not extend beyond the requirements of the Specification, especially in those cases where the Works to be provided by the Contractor form part of some larger project. This is particularly important where, for example, the Contractor is by the Specification required to produce Works which are to form part of some process plant designed and supplied by others. If the Specification requires a certain level of performance of the Plant and it should subsequently turn out that this level of performance is not adequate to produce the desired result for the larger scheme, the Contractor is not to be liable for the inadequacy unless the Contract otherwise provides. The Contract should only provide otherwise if the Contractor, himself, has determined the crucial requirements of the Specification.

14.1 and 14.2 Programme and Form of programme
It is for the Engineer and Purchaser to decide what form the Programme should take, eg. whether it should be in the form of a critical path network or other form of presentation and the extent of detail required. The Programme is important for both Purchaser and Contractor. The Programme should lay down the times by which the Purchaser is to provide drawings, to provide access and services to Site, to obtain any necessary import permits or licences or to complete any civil engineering work and provide design data. If the Purchaser fails to comply with the Programme requirements he is as much in breach of the Contract as would the Contractor be if he failed to comply. The Programme itself may be of considerable use in determining the disrupting effect of any delays which may occur during the course of the Contract and assist the Engineer in determining whether the Contractor is entitled to an extension of time.

14.4 Alterations to programme
In view of the comment on sub-clauses 14.1 (Programme) and 14.2 (Form of programme) it is clear that the Engineer must approve any material alteration to the Programme.

14.5 Revision of programme
This sub-clause gives the Engineer power to order the Contractor to revise the Programme if the progress of the Works either falls behind or moves ahead of the Programme. The intention is to enable the parties to work to a Programme which reflects actual progress rather than the desired progress. The sub-clause further provides that if revision of the Programme is required for

circumstances for which the Contractor is not responsible, eg. where the Purchaser has failed to provide the foundations on time, the Contractor is entitled to be paid the Cost and an allowance for profit of producing the revised Programme. It should be emphasised that the sub-clause gives the Contractor this entitlement only in respect of revising the Programme. It does not deal with or require the Engineer to certify any extra Cost that might be incurred by the Contractor in carrying out work in accordance with the revised Programme. Any entitlement on the Contractor's part to such extra Cost will be governed by the other provisions of the Conditions and in particular by clause 41 (Claims).

14.6 Rate of progress

This is an important provision which empowers the Engineer to give notice to the Contractor if the Engineer decides that the rate of progress of the Works or of any Section is too slow to meet the Time for Completion in circumstances where the Contractor is not entitled to an extension of time. Having received such a notice, the Contractor is required to take such steps, which must be approved by the Engineer, to remedy or mitigate the likely delay including, where necessary, revision of the Programme. The Contractor is not entitled under this sub-clause to payment for taking such steps. However, the Engineer who seeks to operate the provisions of this sub-clause must take the greatest possible care to ensure that he can justify serving the notice, otherwise the Purchaser will almost certainly have to reimburse to the Contractor any "acceleration" Cost that the Contractor may have incurred in consequence of the Engineer's notice.

15.1 Drawings

This sub-clause states the Contractor's obligations in relation to the submission of drawings to the Engineer. For the avoidance of doubt it should be made clear that information to be provided by the Contractor could include calculations. The sub-clause provides that if the Engineer fails to give his approval or disapproval of drawings, etc. within 30 days after they have been submitted (if there is no other time limit for approval or disapproval set out in the Contract or Programme) the drawings will be deemed to have been approved.

The main purpose of requiring the Engineer to approve drawings, etc. is for the Engineer to check that they are generally consistent with the Specification and interface with equipment, etc. supplied by others for the Purchaser's overall project. It will be noted that by reason of sub-clause 16.1 (Errors in drawings, etc. supplied by contractor) approval by the Engineer of drawings does not relieve the Contractor from responsibility for errors, omissions, or discrepancies therein.

15.2 Consequences of disapproval of drawings

If the Engineer disapproves any drawing it must be modified and resubmitted without delay.

15.3 Approved drawings

Once the drawings have been approved by the Engineer they may not be departed from either by the Contractor or by the Engineer. If changes are required which do not involve errors, omissions or discrepancies therein, these should be the subject of a variation under clause 27 (Variations).

15.4 Inspection of drawings

This is purely an administrative procedure for the Engineer's benefit making it clear that he has the right to inspect drawings of any part of the Works at all reasonable times.

15.5 Foundation, etc. drawings

This sub-clause deals with the provision by the Contractor of certain vital information that the Engineer and Purchaser may require in order to enable civil engineering work to be designed and constructed by others, eg. foundations and means of support, the details of any access required and of necessary connections to the Plant such as power and utilities. Timely supply of the information is usually crucial to enable the Works to be completed in time, and it may be appropriate for the Purchaser to make some additional provision in the Special Conditions for liquidated damages for delay should, for reasons within the Contractor's control, drawings and information not be supplied by the due date.

15.6 Operating and maintenance instructions

This sub-clause imposes a positive duty on the Contractor to supply operating and maintenance instructions in addition to as-built drawings of the Works. It should be noted that by reason of the last paragraph of this sub-clause, even if the Works are otherwise complete for the purposes of taking-over by the due date, the absence of the operating and maintenance instructions and drawings entitles the Engineer to withhold the issue of the Taking-Over Certificate and may well entitle the Purchaser to damages for delay. The Purchaser is entitled after all to know how to run and maintain the Works when he first puts them to use.

15.7 Purchaser's use of drawings, etc. supplied by contractor

This sub-clause limits the extent to which the Purchaser and the Engineer may use drawings and information supplied by the Contractor. It should be noted that, in particular, whilst the Purchaser is entitled to use them for the purposes of completing, maintaining, and adjusting and repairing the Works he is not entitled to use them for the purpose of making or having spare parts made by others for the Works. If the Purchaser requires drawings and information for this purpose, this requirement should be made known at the tender stage in order that the Contractor may determine whether to grant a licence for the purpose and, if so, upon what terms.

15.8 Contractor's use of drawings, etc. supplied by purchaser or engineer

This sub-clause restricts the Contractor's use of drawings and information supplied by the Purchaser or the Engineer to such use as is necessary for the purposes of the Contract.

15.9 Manufacturing drawings, etc.

This sub-clause reflects normal practice in the electrical and mechanical engineering plant industry which is not to provide shop or manufacturing drawings. Where it is appropriate for such drawings to be provided, then the obligation to do so should be specified in detail in the Special Conditions and this sub-clause must be deleted in its entirety.

16.1 Errors in drawings, etc. supplied by contractor

The Contractor is responsible for errors, omissions or discrepancies in the drawings, notwithstanding that they may have been approved by the Engineer, unless this is due to incorrect drawings, etc. supplied by the Purchaser or the Engineer. The Contractor must bear the cost of correcting any errors and, at his own expense, carry out the necessary remedial work and modify the drawings accordingly. The purpose of the last paragraph of the sub-clause is to make it clear that, except for any liability for delay that may be caused by the carrying out of the necessary remedial work and modifications to the drawings, the Contractor has no further liability in this respect.

16.2 Errors in drawings, etc. supplied by purchaser or engineer

This sub-clause makes the Purchaser responsible for any errors, etc. in drawings and information supplied by him or the Engineer and requires the Purchaser to carry out the necessary remedial work or to pay the Contractor for so doing (including an allowance for profit). By reason of the provision of sub-clause 44.4 (Exclusive remedies) this is the Contractor's only remedy against the Purchaser in respect of such errors.

17.1, 17.2 and 17.3 Contractor's representatives and workmen, Objection to representatives and Returns of labour

These three sub-clauses deal in general terms with the need for the Contractor to be represented on Site by one or more competent representatives. The Contractor should notify the Engineer of his or their identity before work on Site commences. Instructions given by the Engineer to such a representative relating to work on Site will be deemed to have been given to the Contractor. Whilst the Contractor is responsible for management of his own labour on Site, it must be open to the Purchaser through the Engineer to object to any of the Contractor's employees who are incompetent or otherwise misconduct themselves on Site and sub-clause 17.2 so provides.

Sub-clause 17.3 which deals with returns of labour assists the Engineer in checking claims for additional payment under the Contract and generally to ascertain the nature of work being undertaken and the general level of effort on Site.

18.1 Fencing, guarding, lighting and watching

This sub-clause sets out general regulations for safety precautions to be taken on Site. The general obligations detailed in sub-clause 18.1 may well be supplemented by the Purchaser's own site safety regulations, copies of which should be available to the Contractor at the tender stage in order that he may allow for any extra cost of compliance in his prices. Equally, if a "permit to work" system is in operation at the Site, full details should be given at the tender stage and the Contractor's obligations in relation thereto should be specified in the Special Conditions.

18.2 Site services

This sub-clause is the corollary to sub-clause 11.6 (Utilities and power) and deals with the use by the Contractor of any services that may have been made available by the Purchaser. Ideally, rates for such use should be set out in the Special Conditions and agreed at the tender stage. If they cannot be so agreed, then the Engineer has power under this sub-clause to fix reasonable rates for such use.

18.3 Clearance of site

This sub-clause imposes on the Contractor a general obligation to practise good housekeeping during the progress of the Works on Site and generally to leave the Works clean and in a safe and workmanlike condition to the Engineer's reasonable satisfaction on completion.

18.4 Opportunities for other contractors

The importance of this sub-clause on a multi-discipline Site will be only too obvious, particularly where the Works form only a part of some larger project of the Purchaser. It will be particularly important to ensure that work being done by several contractors on the Site at the same time does not cause disruption to the regular progress of the Contractor's work or to that of others, otherwise the Purchaser may find himself faced with a large number of claims for extra cost from all those engaged on the Site because an inadequate attempt has been made to integrate their respective contributions to the Purchaser's project. It will of course be a necessary consequence of the work being carried out on a multi-discipline Site that the Contractor must provide reasonable opportunities to other contractors to carry out their work.

19.1 and 19.2 Hours of work and No night or rest day working

Sub-clause 19.1 gives the Engineer power to order that work be done outside normal working hours. If the need for this has arisen through no fault of the Contractor, the Contractor is entitled to recover the extra Cost of work so done.

Sub-clause 19.2 recognises that, in general, work is not to be done outside normal working hours or on locally recognised days of rest unless some emergency arises or the Engineer gives his consent. The Engineer is not to withhold his consent if the Contractor requests permission to work outside normal working hours in order to make up time lost.

20. Safety

This clause sets out general obligations of the Contractor in relation to safety and compliance with the Purchaser's safety regulations applicable at the Site. This will not of course override or exclude the need for the Contractor to comply with any applicable legislation relating to safety at work.

21.1, 21.2, 21.3 and 21.4 Extraordinary traffic, Special loads, Extraordinary traffic claims and Waterborne traffic

In the United Kingdom, at least, the matters covered by these sub-clauses in the case of public roads will be covered by the provisions of the Highways Act, 1959. That Act empowers a highway authority to recover the cost of repairs to a highway because of damage caused by excessive weight on it or other extraordinary traffic. The highway authority can recover expenses, incurred as a result of damage arising from the extraordinary traffic, from the person or body by, or in consequence of, whose order that traffic has run.

The Contractor may not be permitted to carry out the special protection or strengthening that he proposes. If is known at the tender stage by reason of the information given by the Contractor to the Engineer as to the projected sizes of loads he will need to deliver to Site and it is established that the Contractor will not himself be able to carry out the necessary protective or strengthening work, then appropriate provision should be included in the Special Conditions if required so as to shift the cost of the necessary work on to the Contractor.

Any Cost incurred by the Contractor because of work done under sub-clause 21.2 shall, together with a profit element, be added to the Contract Price.

The purpose of sub-clause 21.3 is to make it clear that in so far as the Contractor does take reasonable care to prevent damage to highways and bridges, etc., it is the Purchaser's responsibility, via the Engineer on his behalf, to negotiate any extraordinary traffic claims and settle them.

Sub-clause 21.4 applies similar provisions where transport is by waterways.

22.1   Setting out
It is the Contractor's obligation to set out the Works accurately in relation to the relevant points, lines and levels that have been given to him by the Engineer. The Contractor must at his own expense correct any errors that appear in the execution of the Works unless this results from incorrect information supplied in writing by the Purchaser or the Engineer or the Engineer's Representative or by reason of some default on the part of another contractor. Where extra Cost is incurred by the Contractor for reasons attributable to incorrect information supplied in writing by the Purchaser or the Engineer, etc. such Cost, together with a profit element, shall be added to the Contract Price. Although the clause does not specifically so state, the fact that the Engineer may have checked the Contractor's setting out will not relieve the Contractor from responsibility for accurately setting out the Works.

23-24. INSPECTION AND TESTING OF PLANT BEFORE DELIVERY

23.1   Engineer's entitlement to test
The Engineer's right to inspect, examine and test is of general application and includes the right to check the progress of manufacture and in so doing he will, no doubt, compare actual progress with planned progress as shown in the Programme for the purposes of sub-clause 14.6 (Rate of progress). The Engineer's rights are not confined to carrying out these activities at the Contractor's premises since the sub-clause provides that, if Plant is being manufactured elsewhere, the Contractor is obliged to obtain permission for the Engineer to inspect that Plant at its actual place of manufacture. Where the Plant is being manufactured by a Sub-Contractor, the Contractor will need to ensure that provision is included in the sub-contract so as to enable the Contractor to comply with the obligations under this sub-clause. This will usually be done by stipulating a "back-to-back" requirement. Use of the Form of Sub-Contract published with MF/1 will satisfy this requirement.

23.2   Date for test or inspection
This sub-clause deals only with those cases where the test or inspection required is stipulated in the Contract. Nevertheless the time and place of the test must be agreed. Once the time and place have been agreed the Engineer can attend the tests by giving the Contractor 24 hours' notice. If he does not attend on the agreed date, the Contractor can proceed with the tests and they are deemed to have been made in his presence. Whether the Engineer attends, or not, the Contractor is obliged to forward to the Engineer duly certified copies of the results of the test or the inspection.

23.3   Services for test or inspection
This sub-clause obliges the Contractor to provide the necessary facilities and assistance for any tests or inspections as specified in the Contract.

23.4   Certificate of test or inspection
This sub-clause requires an Engineer who is satisfied with the test or inspection to issue a certificate to that effect.

23.5   Failure on test or inspection
If as a result of the test or inspection, the Engineer decides the Plant is defective or not in accordance with the Contract, he can reject it. He must give notice within 14 days of his decision with reasons. Following rejection the Contractor must make good or otherwise repair or replace the rejected Plant and resubmit it for test or inspection. The Purchaser is entitled to recover all expenses reasonably incurred in connection with any such re-testing or re-inspection.

24.1   Delivery

The Contractor must have written permission from the Engineer to deliver any Plant or Contractor's Equipment to the Site. The Engineer has control over deliveries to Site and is entitled to co-ordinate these in a fair and reasonable manner having regard to the Programme or other provisions of the Contract. If the Engineer fails to give the necessary permission to deliver, the Contractor may be entitled to treat such a failure as instructions to suspend delivery in which event the provisions of clause 25 (Suspension of work, delivery or erection) -see below- will apply. The Contractor is responsible for the reception and unloading of all Plant and Contractor's Equipment on Site.

25.   SUSPENSION OF WORK, DELIVERY OR ERECTION

25.1   Instructions to suspend

The Engineer has power to suspend the progress of the Works at any time. The Engineer will be deemed to have given instructions to suspend the progress of the Works, if the Contractor is prevented from delivering Plant to Site at the time for delivery specified in the Programme, or at what would otherwise be the appropriate time, or is prevented from erecting any Plant because of some delay or failure on the part of the Engineer to give permission to deliver or by reason of any cause for which the Purchaser or some other contractor employed by him is responsible. If instructions to suspend are given or deemed to have been given, then the Contractor must store the Plant and preserve, protect and otherwise secure the Works, as the case may be, and insure them to the extent that the Engineer may require. Finally, the sub-clause makes it clear that although the Engineer may have given instructions to suspend progress, this of itself will not entitle the Contractor to withdraw his staff, labour and Contractor's Equipment from Site unless the Engineer has instructed him to do so.

25.2   Additional cost caused by suspension

The Contractor is entitled to be paid the additional Cost and profit, under sub-clause 41.2 (Allowance for profit on claims), of complying with the Engineer's instructions under sub-clause 25.1 unless sub-clause 25.4 applies (see below) as long as the Contractor makes his claim in accordance with clause 41 (Claims).

25.3   Payment for plant affected by suspension

If the Contractor is instructed to suspend delivery, then, depending upon the terms of payment, the Contractor may well find himself in a position where he could not obtain the payment due on delivery because the Plant has, in fact, not been delivered to Site. If the suspension lasts for longer than 30 days this sub-clause entitles the Contractor to apply for the payment which would have been due on delivery had delivery been possible. Again, the Contractor's entitlement to such payment is subject to sub-clause 25.4 (see below). Additionally, however, in order to obtain payment under this sub-clause, the Contractor must mark the Plant as the Purchaser's property under sub-clause 37.2 (Marking of plant) and insure it as if it were on Site.

25.4   Disallowance of additional cost or payment

Although under sub-clause 25.2 the Contractor is prima facie entitled to recover the additional Cost and profit for complying with the Engineer's instructions to suspend under sub-clause 25.1, he is not entitled to be paid any such additional Cost or profit if the suspension was necessary because of the Contractor's default or was required for the proper execution or safety of the Works. If the necessity for suspension to ensure for the proper execution or safety of the Works has occurred in consequence of an act or default of the Engineer or the Purchaser or by reason of any of the Purchaser's Risks (see clause 45) the Contractor is entitled to recover the additional Cost and profit.

25.5   Resumption of work, delivery or erection

This sub-clause sets out the procedure for dealing with a resumption of work following suspension. It further provides that if instructions to proceed have not been given within 90 days after the order to suspend, the Contractor can give notice to the Engineer requiring him to give notice to proceed within 30 days. If the Engineer does not give that notice then the Contractor can either treat the suspension as an omission of the Section of the Works that is affected by the order to suspend, or, if it affects the whole of the Works, he can terminate the Contract. In the latter event the Contractor is entitled to be paid as if the Contract has been terminated by the Contractor because of the Purchaser's default [see sub-clauses 51.1 (Notice of termination due to purchaser's default) and 51.3

(Payment on termination due to purchaser's default)]. If these alternatives are not appropriate and the Contractor is prepared to wait for instructions to proceed, he is entitled to be paid the Contract Value of the Plant affected by the suspension.

When the Contractor is given notice to proceed, he must in all cases examine the Plant and work that has been affected by suspension and report to the Engineer on any deterioration or defect that he finds. If the Plant needs to be made good in consequence, the Contractor is entitled to be paid the Cost for so doing (and, under sub-clause 41.2 (Allowance for profit on claims), a profit element) unless the work was necessary because of defective materials or workmanship or because the Contractor failed to comply with the Engineer's instructions under sub-clause 25.1.

25.6 Effect of suspension on defects liability

This sub-clause entitles the Contractor to recover the additional Cost of complying with his obligations in relation to defects under clause 36 (Defects liability) and, under sub-clause 41.2 (Allowance for profit on claims), a profit element if, in consequence of an order to suspend, the defects come to light more than three years after the Normal Delivery Date.

26. DEFECTS BEFORE TAKING-OVER

26.1 Defects before taking-over

This clause contains a general power for the Engineer at any time before taking-over to reject any work done, Plant supplied or materials used by the Contractor which is defective or is not in accordance with the Contract. If he decides to reject such work or Plant he must notify the Contractor as soon as possible specifying particulars of the defects and the Plant must be placed at the Contractor's disposal. The Contractor is obliged with all speed to make good the defects and if he fails to do so the Purchaser may take steps to make good the defects at the Contractor's expense. If the only way of so doing is for the Purchaser to purchase replacement Plant he may do so at reasonable prices and under competitive conditions where possible and charge the Contractor accordingly. In such case the Contractor is entitled at his own expense to remove and retain any Plant that the Purchaser may have replaced.

It should also be noted that the last sentence of the sub-clause makes it clear that any right the Purchaser may have under the provisions of clause 34 (Delay) to damages for delay is unaffected by the actions a Purchaser may take under the provisions of clause 26.

27. VARIATIONS

27.1 Meaning of variation

It was thought useful to include within this sub-clause a definition of the term "variation". In essence any alteration of the Works as described in the Specification will constitute a variation. See also the comment made for sub-clause 44.4 (Exclusive remedies) below.

27.2 Engineer's power to vary

Any instructions in relation to variations must be given by the Engineer by notice and must therefore be in writing. If an instruction to vary is given orally, the Contractor must for his own protection utilise the provisions of sub-clause 2.5 (Confirmation in writing) to ensure that he has written instructions to carry out the variation. Once written instructions have been received the Contractor must give written notice to the Engineer if in the Contractor's view an addition or deduction from the Contract Price is warranted by reason of the variation.

The Engineer has no power to order variations which would cause an addition to or deduction from the Contract Price of more than 15% without the written consent of both the Purchaser and the Contractor. This places a limit on the value of variations that may be ordered for which the Cost is to be determined in accordance with sub-clause 27.3. It also places a limit on the freedom of the Engineer to authorise variations without first seeking the agreement of the Purchaser.

The sub-clause makes clear that the Contractor may, of course, propose variations to the Engineer at any stage but he may not proceed with any such variation unless he has received written directions to do so from the Engineer.

27.3 Valuation of variations

This sub-clause reflects the fact that in many cases prior to giving formal variation instructions, the Engineer may well have requested the Contractor to provide quotations for the work in question which will have been accepted by the Purchaser before the Engineer gave his instructions. Where this does not apply, the amount to be added to or deducted from the Contract Price is to be determined by the Engineer in accordance with rates specified in the schedule of prices if they are applicable. If the rates are not applicable, then the Contractor is entitled to be paid or must allow to the Purchaser a reasonable sum. Users may consider it appropriate to provide in the Special Conditions for payments on account for variations in those cases where the adjustment cannot be determined within say 3 months after the variation instruction is given provided, of course, that the Contractor has taken all reasonable steps to provide quotations where this procedure is applicable.

27.4 Contractor's records of costs

In many cases it may not be possible to establish the adjustment to be made to the Contract Price on account of the variation before instructions to carry out the variation are given. Accordingly, this sub-clause requires the Contractor to keep contemporary records of Cost and time expended on the variation and the Engineer will be entitled to inspect these at all reasonable times. These records will assist the Engineer in determining in appropriate cases the amount of any "reasonable sum" to be added to or deducted from the Contract Price under sub-clause 27.3.

27.5 Notice and confirmation of variations

The Engineer must give the Contractor reasonable notice of any variations he may require. If, in consequence of the variations, work already done must be altered, the Contractor is entitled to be paid the Cost of the alterations. Under sub-clause 41.2 (Allowance for profit on claims) the Contractor may also claim a profit element.

In addition to the general effects of the variation on work already done or to be done, it may be that, in consequence of a variation, the Contractor may also consider that he would not be able to fulfil obligations under the Contract to which he may previously have been committed. For example, he may consider that, as a consequence of the variation, the performance of the Works, as originally envisaged, may be affected. In such circumstances the Contractor must notify the Engineer, with full supporting details, and the Engineer must decide whether the variation is to be carried out and until the Engineer gives his instructions on the matter the variation will be deemed not to have been given and the Contractor is not, therefore, obliged to proceed with it. If the Engineer confirms his instructions in relation to the variation he should, to the extent that may be justified, modify the obligations which the Contractor considers will be affected.

27.6 Progress with variations

On receipt of instructions under sub-clause 27.2 or confirmation of instructions under sub-clause 27.5 the Contractor is to proceed with the variation ordered unless he has already notified the Engineer that an adjustment of more than 15% of the Contract Price will be warranted. Only in those circumstances or in the circumstances envisaged by sub-clause 27.5 may the Contractor delay carrying out the variation instructions pending agreement on price.

28. TESTS ON COMPLETION

This clause deals only with the procedures to be followed in carrying out the Tests on Completion. It does not deal with the details of the tests that need to be passed. It is normal for details of the specific tests to be passed to be included in the Specification or if it is not possible to specify the tests at the date of the Contract (this may particularly not be possible where the tests involve Tests on Completion of Software) the agreed tests should be incorporated in a schedule, signed by the parties and annexed to the Contract. Such tests should be the subject of a variation order (but the Contractor will not necessarily be entitled to extra Cost).

28.1 Notice of tests

The Contractor must notify the Engineer when the Works are ready for the Tests on Completion. 21 days' advance notice must be given. The Engineer and Contractor must agree on a specific date for the tests but if they do not do so they must take place or at least commence within 10 days after the 21-day period has expired and the Engineer is entitled to notify the Contractor of the date when he wants the tests to be undertaken.

28.2   Time for tests

If either the Engineer fails to appoint a time for the tests or having done so does not attend, in either instance the Contractor is entitled to proceed with the tests in the Engineer's absence. Despite this, it would be sensible, if the Engineer has failed to appoint a time, for the Contractor to notify the Engineer of the anticipated date and place of the tests, even though he is under no obligation under the sub-clause to do so.

If the tests are carried out by the Contractor in the Engineer's absence he must send duly certified copies of the results of the Tests on Completion to the Engineer.

28.3   Delayed tests

This sub-clause deals with those cases where Tests on Completion are being unduly delayed by the Contractor. This will normally only occur if the Contractor has already exceeded the Time for Completion. If these circumstances arise, the Engineer can require the Contractor to carry out the tests within 21 days after receipt of the notice from the Engineer. If the Contractor receives such a notice he can fix the date and time of the tests but must notify the Engineer accordingly. If the Contractor fails to make the tests then the Engineer is entitled to proceed with them at the Contractor's risk and expense and to deduct the Cost from the Contract Price. However, if it subsequently transpires that the tests were not being unduly delayed by the Contractor, the sub-clause provides that any tests that the Engineer may have carried out will be made at the risk and expense of the Purchaser. Whilst, therefore, the provisions of this sub-clause provide a useful remedy in relation to the Tests on Completion in a case where the Contractor is unduly dilatory, the Engineer must be sure of his ground before proceeding to test the Works himself.

28.4   Repeat tests

This sub-clause permits the Contractor to repeat the tests within a reasonable time if the Works fail to pass the Tests on Completion on the first occasion. If the Purchaser incurs extra Cost in this connection, for example, by providing additional feedstock, or simply the provision of the necessary power and services, the Cost is to be deducted from the Contract Price.

28.5   Consequences of failure to pass tests on completion

This sub-clause makes it clear that if the Works fail to pass the Tests on Completion, whether this be on the first occasion or upon a repetition under sub-clause 28.4, the Contractor is to do whatever is necessary to enable the Works or the relevant Section to pass the tests, which must then be repeated unless a specific time limit has been specified in the Contract for passing the tests. If that time limit has been exceeded, the Engineer is entitled to reject the Works and treat the Contractor as in default under clause 49 (Contractor's default).

29-31. TAKING-OVER

Taking-over is essentially the end of the construction phase of the Works or Section, as the case may be, when the Tests on Completion have been passed and the Works or Section are complete in all material respects and are accepted by the Purchaser. At this moment the risk of loss or damage to the Works or Section will pass to the Purchaser and the Contractor's obligation in relation to defects will commence. The date certified in the Taking-Over Certificate issued on taking-over will, subject to any right the Contractor may have to an extension of time under the provisions of clause 33, determine the extent to which the Contractor may be liable for delay.

29.1   Taking-over by sections

This sub-clause makes it clear that if the Contract provides for separate taking-over of Sections of the Works the clause is equally applicable.

29.2   Taking-over certificate

This sub-clause makes it clear that it is the Engineer's duty, when the Works have passed the Tests on Completion and are complete to the extent that they can be used for their intended purpose, to issue a Taking-Over Certificate. That certificate when issued must certify the date on which the Works passed the Tests on Completion and were so complete. It is from that date and not the date of the Taking-Over Certificate that the Contractor's obligations in relation to defects under clause 36 (Defects liability) will commence. Subject to the provisions of clause 30 (Use before taking-over) (as to which see comments below) the Purchaser is not permitted to use the Works before they are taken over.

29.3 Effect of taking-over certificate

From the date stated in the Taking-Over Certificate, ie. upon successful completion of the taking-over tests for the Works or a Section, the risk of loss of or damage to the Works or the Section will pass to the Purchaser. The Contractor's "Contractors All Risks" insurance will cease 14 days after the date of issue of the Taking-Over Certificate and the Engineer must therefore inform the Purchaser of his intention to issue a Taking-Over Certificate in sufficient time to enable the Purchaser to arrange his own insurance cover.

29.4 Outstanding work

Even though the Works may have passed the Tests on Completion and can be used by the Purchaser for the purpose for which they are intended, there may well be defects of a minor character or outstanding items detailed in the Taking-Over Certificate which the Contractor is still required to correct or complete. The Form of Taking-Over Certificate included with the model form provides that a list of these items shall be annexed to the certificate and instructs the Contractor to complete or correct them within 28 days. If this Form of Taking-Over Certificate is used by the Engineer, and it would be reasonable for the work to be completed and the defects corrected within that 28 day period and the Contractor fails to do so, or if the Contractor fails to do so within a reasonable time, the Purchaser can arrange for the outstanding work to be done by others, in which event the Cost of that work is to be certified by the Engineer and deducted from the Contract Price.

30.1 Use before taking-over

Under the provisions of sub-clause 29.2 (Taking-over certificate) the Purchaser is not entitled to use the Works or any Section unless a Taking-Over Certificate has been issued. However, a Taking-Over Certificate may not have been issued for the whole of the Works within one month after the Time for Completion because of reasons for which the Contractor is responsible. In such circumstances, the Purchaser is entitled to use those Sections or parts of the Works which are reasonably capable of being used. However, any such use must be at the Purchaser's risk and his insurance arrangements should be made accordingly. Additionally, if the Purchaser does use a Section or part of the Works in these circumstances, the Defects Liability Period will commence on the date the relevant Section or part is taken into use. Notwithstanding that the Purchaser may be using the Works under the provisions of this clause, the Contractor is to be given the earliest possible opportunity to enable him to take the necessary steps to permit the issue of the Taking-Over Certificate by, for example, completing any outstanding Tests on Completion.

31.1 Interference with tests

This sub-clause applies to those circumstances where by reason of some default on the part of the Purchaser, the Engineer, or other contractors engaged by the Purchaser, the Contractor is prevented from carrying out the Tests on Completion. In such circumstances the Engineer has a duty to issue a Taking-Over Certificate notwithstanding that the tests have not been passed unless he can show that the Works are not substantially in accordance with the Contract.

31.2 Tests during defects liability period

If the Engineer has been obliged to issue a Taking-Over Certificate under sub-clause 31.1 the Contractor is not released from any obligation to carry out the Tests on Completion, indeed he is under a duty to carry them out during the Defects Liability Period. However, he is entitled under those circumstances to recover any additional Cost incurred (and, under sub-clause 41.2 (Allowance for profit on claims), a profit element) because the tests were made at a later date than that envisaged. The Engineer must give the Contractor 14 days' notice of when he requires the Tests on Completion to be carried out in these circumstances.

32 – 33. TIME FOR COMPLETION

32.1 Time for completion

If the Contractor fails to complete as required by this clause, then he may either be entitled to an extension of time under clause 33 or may be liable to the Purchaser for delay under the provisions of clause 34 (delay). It should be emphasised that under the model form the expression "completion" means the bringing of the Works by the Contractor to the stage where the Tests on Completion have been passed and the Works can be put to their intended use. Time for Completion does not therefore encompass the passing of the Performance Tests. The reason for this is that the Contractor has no direct control over the time at which Performance Tests are carried out and indeed these tests are to be carried out by the Purchaser or the Engineer under the Contractor's supervision and the timing would not necessarily be fixed by the Contractor.

If separate Times for Completion and separate damages for delay are fixed for Sections of the Works, the appropriate additional Special Conditions may be used (see also section 5 of this Commentary).

33.1 Extension of time for completion

If variation orders and acts or omissions on the part of the Purchaser or the Engineer were not included as specific grounds for extension of time, neither party would, should the Contractor be late, be entitled to rely upon the provisions for liquidated damages contained in sub-clause 34.1 (Delay in completion), with the consequences that damages for delay would be at large and the Purchaser would have to prove his loss. In this connection it should be remembered that the making of a variation order at a time when liquidated damages are accruing under sub-clause 34.1 may have the effect of destroying the Purchaser's accrued rights to liquidated damages.

Industrial dispute is included as a specific ground for extension of time in order to prevent the Engineer or, ultimately, an arbitrator in any dispute under the Contract from enquiring into the cause of the Contractor's labour problems. All the Contractor has to show is that there was an industrial dispute, whether amongst his own workmen or unconnected with the Site, the effects of which were to delay or impede him in performance. The Engineer must establish the facts and then determine upon an objective basis the amount of any extension of time that may be justified in the circumstances. Apart from these particular circumstances, the sub-clause leaves it open to the Engineer to decide on the facts of each case whether the Contractor is entitled to an extension of time. It should be noted that an extension of time may be applied for and granted except in a case of prolonged delay to which sub-clause 34.2 (Prolonged delay) applies, even though the contractual Time for Completion has already been exceeded, provided that the Contractor has in all cases as soon as reasonably practicable given the Purchaser or the Engineer notice of his claim with full supporting details.

33.2 Delays by sub-contractors

This sub-clause removes any argument there might be as to the extent to which delays by Sub-Contractors would entitle the Contractor to an extension of time. The Contractor will be entitled to an extension only if the circumstances which caused the Sub-Contractor to be delayed would constitute valid grounds for the Contractor to claim an extension of time, had those circumstances been suffered by him.

33.3 Mitigation of consequences of delay

The purpose of this sub-clause is to impose a positive duty on the Contractor and the Engineer, in circumstances where the Contractor has claimed an extension of time, to review all the circumstances and determine whether any steps can be taken to overcome or reduce the actual or anticipated delay. The Contractor must comply with any instructions in this respect that the Engineer may give and if he incurs additional Cost in so doing and if he is entitled to an extension of time, such additional Cost and, under sub-clause 41.2 (Allowance for profit on claims), a profit element, will be added to the Contract Price. If the Contractor was not entitled to claim the extension of time but, nevertheless, the Engineer gave instructions to mitigate the consequences of delay, it would seem more appropriate for the Engineer to give the necessary instructions not under the provisions of this sub-clause but under the provision of sub-clause 14.6 (Rate of progress) in order to avoid any argument that the Contractor was entitled to be paid extra Cost.

34. DELAY

The general practice of the electrical and mechanical engineering plant industry of offering liquidated damages calculated by reference to a percentage of the Contract Value of those parts of the Works which cannot be put to their intended use for each week of delay up to a maximum is maintained. However, a provision has been introduced to deal with circumstances where the Purchaser is already entitled to the maximum damages for the delay, yet the Works are still incomplete. Since it is considered that in electrical and mechanical engineering plant contracts which include installation, erection and testing, the Time for Completion is not of the essence unless specifically so provided, sub-clause 34.2 provides that once the maximum liquidated damages has been reached the Purchaser may fix a final Time for Completion which must in all the circumstances be reasonable. Certain consequences follow should the Contractor still fail to complete after such a notice has been given.

34.1 Delay in completion

Completion for the purposes of this sub-clause does not include the Performance Tests nor fulfilment of the Contractor's obligations in relation to defects. The time by which the Contractor must complete is accordingly the time at which the Works are taken over. If no Time for Completion is given in the Programme or in the Contract the Contractor's obligation is to complete within a reasonable time. On the expiration of such a reasonable time liquidated damages will start to run. It is of the essence of this sub-clause that the Purchaser cannot treat the Contract as having been repudiated merely by reason of either the agreed Time for Completion (plus any extensions) or a reasonable time having been exceeded. The Purchaser is no longer required to show that he has suffered any loss (however nominal) from the Contractor's failure to complete. It is generally accepted that those words nowadays fulfil very little purpose since if a Purchaser were able to show that he had suffered some loss (however small) that was more than nominal, the requirement would be satisfied. It was difficult to conceive of any contract for electrical and mechanical work where given delay in completion a Purchaser could not prove that he had suffered more than nominal loss.

Unlike building contracts where damages payable are normally a sum of money per day or per week with no limit, the rate of liquidated damages is fixed by reference to percentages of the Contract Value of those parts of the Works that cannot be used in consequence of the delay. Subject to the provisions of sub-clause 34.2 the maximum damages are in full satisfaction of the Contractor's failure to complete on the contractual date. Sub-clause 34.2 deals with the position where the Purchaser's remedy in liquidated damages under sub-clause 34.1 becomes exhausted. The rates of liquidated damages commonly provided vary between 1/4% and 1% per week and provide for maximum damages of between 5% and 15%. In every case where the parties intend that liquidated damages shall be applied for late completion the rates must be agreed pre-Contract if the provisions are to be enforceable. It is of course open to the parties to agree a specific sum per day or week in which case an appropriate amendment to the sub-clause will have to be included in the Special Conditions.

Where the parties intend that Sections of the Works are to be taken-over separately, the Time for Completion of each Section and the liquidated damages, if any, to be applied in the event of delayed completion of each Section must be defined in the Appendix to the general conditions.

34.2 Prolonged delay

This sub-clause makes it clear that where the Purchaser's remedy in liquidated damages is exhausted the Contractor may not sit idly by and take his time. Once the maximum liquidated damages has been reached, the Purchaser can immediately give notice to the Contractor fixing a final Time for Completion which is reasonable having regard to the delay that has already occurred and to the extent of the work required for completion. This reflects what would be the common law position. If, for any reason whatsoever other than one for which the Purchaser or some other contractor employed by him is responsible, the Contractor fails to meet the final time, the Purchaser has two different remedies open to him.

First, and of course this will be most useful where very little remains to be done to complete the Works, he may again require the Contractor to complete.

Second, he has the option to terminate the Contract in respect of those parts of the Works that remain incomplete. It is submitted that if the part of the Works remaining incomplete does not permit the completed parts of the Works to be used for his intended purpose, the Purchaser would be entitled to terminate the Contract in respect of the whole of the Works.

Whatever remedy the Purchaser chooses to apply he can also recover from the Contractor any loss he has suffered because of the Contractor's failure to comply with the final Time for Completion up to an agreed limit to be stated in the Appendix or, if no such limit is stated, up to the amount of the Contract Value of those parts of the Works that cannot be used by reason of the failure.

35.   PERFORMANCE TESTS
These provisions follow closely the provisions relating to performance tests which are included in the Institution of Chemical Engineers' model forms of contract for process plant.

35.1   Time for performance tests
Not every Contract subject to the model form will include provisions for Performance Tests. The time for Performance Tests may be included in the Contract but if not they are to be carried out as soon as reasonably practicable and within a reasonable time after the Works have been taken over.

35.2   Procedures for performance tests
The Performance Tests are to be carried out by the Purchaser or the Engineer under the supervision of the Contractor and in accordance with procedures and operating conditions specified in the Contract. The Contractor is entitled to give instructions in the course of carrying out the tests and the Purchaser or the Engineer as appropriate must comply with them.

35.3   Cessation of performance tests
At any time whilst the Performance Tests are being carried out either the Purchaser, Engineer or Contractor can order them to cease if their continuation is likely to result in damage to the Works or personal injury.

35.4   Adjustments and modifications
If there is a failure to pass the Performance Tests then the Contractor must be permitted by the Purchaser to make adjustments and modifications before any Performance Tests are repeated. The Contractor can also require the Purchaser to shut down any part of the Works so that he may make any necessary adjustments and modifications and order the Purchaser to restart thereafter. The Contractor must bear the additional Cost incurred by the Purchaser in any repetition of the tests and the expense of any adjustments and modifications. Any proposals for adjustments or modifications must be submitted to the Engineer if the Engineer so requires.

35.5   Postponement of adjustments and modifications
Notwithstanding that the Performance Tests may have failed and that the Contractor wants to make adjustments and modifications, it may not suit the Purchaser to permit the Contractor to make the necessary adjustments and modifications and repeat the tests at a time suitable to the Contractor. This sub-clause permits the Engineer to postpone the adjustments and modifications until such time as he requires them to be carried out by reasonable notice to the Contractor. If, however, the notice requiring the Performance Tests to be performed is not given by the Engineer within one year after the date of taking-over, the Contractor is relieved of any further obligation to make the adjustments or modifications and the Works or relevant Section will be deemed to have passed the Performance Tests.

35.6   Time for completion of performance tests
If a fixed time was given in the Contract for completion of the Performance Tests the Purchaser is entitled to use the Works as he thinks fit once that time has been exceeded.

35.7   Evaluation of results of performance tests
This sub-clause requires the Purchaser, Engineer and Contractor jointly to evaluate the results and it is assumed that the method and manner of evaluation will have been specified in the Contract. In a simple case all that may be necessary may be to determine whether the performance of the Plant falls within stipulated acceptance limits. Where appropriate, allowance must be made in considering the performances achieved for any previous use of the Works by the Purchaser and to differences between operating conditions under which the tests were conducted and those detailed in the Specification or in any performance tests schedule. Generally, the sub-clause requires the results of Performance Tests to be evaluated in an objective manner.

35.8   Consequences of failure to pass performance tests
This sub-clause provides the remedies available to the Purchaser should the Plant fail to pass the Performance Tests. Where performance has been guaranteed and liquidated damages for failure to achieve them have been specified in the Special Conditions and the results are within stipulated acceptance limits, the Contractor must pay or allow the Purchaser the liquidated damages following

which the Purchaser must accept the Works. The acceptance limits should be stated in the Special Conditions.

If guaranteed performances have been specified but the results are outside the acceptance limits, or if there is no provision for liquidated damages, the Purchaser can accept the Works subject to a reasonable reduction in the Contract Price to be agreed by the Purchaser and Contractor or fixed by arbitration.

If, however, the failure of the Performance Tests is such as to deprive the Purchaser of the whole of the benefit of the Works the Purchaser may exercise the ultimate remedy and reject the Works and proceed in accordance with clause 49 (Contractor's default). In such circumstances it would appear that the Engineer in valuing the Works as at the date of termination under the provisions of sub-clause 49.2 (Valuation at date of termination) might be entitled to consider the value to be so small so as to entitle the Purchaser to the recovery of a large part of the Contract Price and to the extra cost of employing another contractor to complete the Works (if that course of action is feasible).

36. DEFECTS LIABILITY

The philosophy behind this clause is that once a Section of the Works has been taken over, the Contractor is responsible for making good any defects which are attributable to his fault and which occur within a defined time (the Defects Liability Period) after taking-over. The Contractor is also liable to make good damage to the Works caused by defects but, that apart, the Contractor is not responsible for any damage to other property of the Purchaser caused by such defects.

Other properties of the Purchaser would generally be covered by the Purchaser's own insurance arrangements and thus double insurance is avoided. It should be noted that under sub-clause 47.2 (Extension of works insurance) the Contractor is required to insure against loss or damage for which he is responsible under clause 36 during the Defects Liability Period.

36.1 Defects after taking-over

It is open to the parties to choose their own Defects Liability Period of whatever length may be appropriate. If they do not do so then the Defects Liability Period will be twelve months from the date of taking-over the Works or, if a Taking-Over Certificate is issued for any Section or part of the Works, the Defects Liability Period for such Section or part is to commence on its separate date of taking-over.

36.2 Making good defects

This sub-clause defines the nature of the defects or damage which the Contractor undertakes to repair or replace. Defective materials, workmanship or design covers the whole area of the Contractor's responsibility. Thus there is no need to make the Contractor's obligations subject to "proper use" of the Works by the Purchaser or to mention acts or omissions of the Purchaser or the Engineer since they are quite clearly not covered by the provisions of the sub-clause. Additionally the Contractor takes responsibility for his acts or omissions during the Defects Liability Period. Thus if the Contractor damages a portion of the Works which has been taken over while attempting to remedy a defect in it, he is liable to make good the resulting damage. Likewise if the Contractor fails to take reasonable precautions to protect a part of the Works not yet taken over and in consequence of repairs to Plant taken over the former part is damaged, the Contractor will be responsible for making good that damage. Finally it is made clear that if the Contractor has been required to accept designs furnished or specified by the Purchaser or the Engineer he will not be responsible for defects in such designs or for any damage to any part of the Works attributable thereto if he had disclaimed responsibility for them in accordance with sub-clause 13.3 (Contractor's design). If the Contractor raises no objection to designs provided by the Purchaser or the Engineer he will be responsible for all the consequences of any defects therein.

36.3 Notice of defects

If defects or damage occur the Purchaser or the Engineer must give immediate notice to the Contractor thereof. Once the defect or damage has been made good then the part repaired or replaced has an additional Defects Liability Period commencing when the work was completed subject, however, to that period not extending beyond two years from the date of taking-over. It is open to the parties to extend this period by agreement and if they so wish to do so the period chosen should be stated in the Special Conditions.

36.4   Extension of defects liability

The nature of some defects or damage may require the whole Works or a particular Section to be shut down. The Defects Liability Period for the parts which cannot be used (but which do not need to be repaired) is accordingly extended for the period of "outage" so that the Purchaser will have the full benefit of the Defects Liability Period.

36.5   Delay in remedying defects

This sub-clause enables a Purchaser to carry out repair work at the Contractor's risk and expense provided he does so in a reasonable manner and notifies the Contractor of his intentions but only where the Contractor has not carried out the repairs within a reasonable time.

36.6   Removal of defective work

The Contractor needs consent to remove a part of the Works which is defective or damaged because it is the Purchaser's property.

36.7   Further tests

This sub-clause permits the Purchaser or the Engineer to require the Works to be re-tested if the repairs or replacements are serious enough to warrant it. Any such re-testing will be at the Contractor's expense.

36.8   Contractor to search

This sub-clause entitles an Engineer who believes that any part of the Works may be defective but who cannot point to any particular defect having manifested itself to require the Contractor to search for the cause. If it is established that the cause is a defect for which the Contractor is responsible then the Cost of the search must be borne by the Contractor but otherwise it must be borne by the Purchaser and, together with, under sub-clause 41.2 (Allowance for profit on claims), a profit element, added to the Contract Price. The Engineer will be entitled to require the Contractor to conduct such a search if, for example, parts of the Works exhibit a deterioration of performance or, say, excessive oil consumption or wear and the cause is not immediately apparent.

36.9   Limitation of liability for defects

This sub-clause makes clear that the Contractor's obligations under this clause, if complied with, must be accepted by the Purchaser in place of any contract terms implied by law as to the quality or fitness for any particular purpose of the Works and in lieu of any liability the Contractor might have in relation to defects caused by the Contractor's negligence or breach of statutory duty. Under English common law a Purchaser has no right to require the remedying of defects by the Contractor once the Works have been accepted. He only has a right to receive damages for the defects in question. The ability to require the Contractor to make good the defects and any damage caused thereby is of considerable advantage to the Purchaser and, further, it should be noted that the Contractor's obligation extends to any defect in design, material and workmanship and not just to defects which would amount to breach of the implied contract terms as to quality, fitness for purpose or workmanship of the Works. This sub-clause will in many cases be subject to the test of reasonableness under the Unfair Contract Terms Act, 1977 and in this connection it is relevant to note that it is far easier and economic for a purchaser to insure against plant breakdown and the consequences thereof than it is for a contractor to do so.

The purpose of the last paragraph of this sub-clause is to make it clear that the limitation of liability is intended also to limit the liability of a Sub-Contractor in a similar manner to that of the Contractor.

Apart from any liability the Contractor may have under sub-clause 36.10 for latent defects (see below) the effect of this sub-clause is to make it clear that clause 36 sets out the limit of his liability and responsibility for defects. It is, however, considered that a Contractor cannot have the benefit of the limitation of liability contained in this sub-clause if he does not, in fact, make good the defects or damage to the Works caused by defects in accordance with the provisions of clause 36.

36.10   Latent defects

This sub-clause gives the Purchaser a remedy for latent defects which appear after the end of the Defects Liability Period and before the expiry of three years after taking-over, in circumstances where the defect was caused by "gross misconduct" as defined in the sub-clause and would not have been disclosed by reasonable examination prior to the expiry of the Defects Liability Period. The definition of "gross misconduct" is intended to confine the Purchaser's remedy to those cases where there has, for example, been deliberate concealment of inferior workmanship or materials or where the Contractor can be shown to have recklessly followed a course of action which no contractor skilled in the relevant art would have followed.

37 – 38. VESTING OF PLANT, AND CONTRACTOR'S EQUiPMENT

37.1   Ownership of plant
The ownership or property in the Plant passes to the Purchaser at whichever is the earlier of the times indicated, regardless of whether or not the risk of loss has passed. Ownership of the Plant will pass either when it is delivered in accordance with the Contract (whether it has been paid for or not) or when the Contractor is entitled to have its value included in an interim certificate of payment if this is earlier. It was not thought appropriate to include in the model form any provision whereby the passing of ownership was dependant upon payment in full or in part.

37.2   Marking of plant
It was considered only fair that where ownership of Plant passes to the Purchaser before delivery to the Site under the provisions of sub-clause 37.1, that the Contractor should be under an obligation, where practicable, to set aside the Plant and physically mark it as the Purchaser's property. It should be noted that the Contract Value of Plant may not be included in any interim certificate of payment to which the Contractor might otherwise be entitled unless and until it has been set aside and marked as required by this sub-clause.

38.1   Contractor's equipment
The purpose of requiring a Contractor to provide the Engineer with a list of the Contractor's Equipment that the Contractor intends to use on the Site is not only so that the Engineer may comment thereon if thought appropriate but also to assist the Engineer in monitoring the safety of operations on Site and in the evaluation of any Contractor's claims for additional payment.

38.2   Contractor's equipment on site
The purpose of this sub-clause is to enable the Engineer to keep himself informed on the use of Contractor's Equipment on Site and its movement to and from the Site.

38.3 and 38.4 Loss or Damage to the Contractor's Equipment and Maintenance of Contractor's Equipment
The object and intent of these sub-clauses needs no further comment or explanation.

39 – 40. CERTIFICATES AND PAYMENT
Although MF/1 contains no model terms of payment, certification procedures remain important in determining the Contractor's entitlement to payment.

39.1   Application for payment
This sub-clause details the items for which the Contractor may make application for interim certificates of payment. It should be noted that these matters may be otherwise regulated in or by the Special Conditions.

39.2   Form of application
The Contractor must make his application in the form of an invoice accompanied by the necessary evidence. Generally Purchasers will have to specify in the Special Conditions the evidence required especially in the case of claims for payment for Plant in the course of manufacture. In the case of Plant delivered, shipped, or en route to the Site the items mentioned in sub-paragraph (b) may be sufficient but the Special Conditions may need to lay down any particular requirements, especially those that may need to be provided to bodies financing the Contract for the Purchaser. However, it should be noted that claims for additional payment must be accompanied by the particulars required under paragraph (b) of sub-clause 41.1 (Notification of claims).

39.3   Issue of payment certificate
The Engineer is required to issue an interim certificate of payment within 14 days after receiving an application therefor which the Contractor was entitled to make. If the application is not

accompanied by the appropriate evidence as required by sub-clause 39.2 the Engineer cannot be required to issue an interim certificate of payment until 14 days after that evidence has been provided.

39.4 Value included in certificates of payment
This sub-clause sets out in detail the manner in which the sum certified as due under the certificate is to be ascertained.

39.5 and 39.6 Adjustments to certificates and Corrections to certificates
No explanation of these administrative provisions is thought necessary.

39.7 Withholding certificate of payment
This sub-clause makes it clear that the Engineer cannot simply withhold a certificate because there are defects of a minor character which would not affect the use of the Works. The Engineer will be perfectly entitled, however, to make allowance for the work that is defective in calculating the total sum to be certified even where the defects are of a minor character.

39.8 Effect of certificates of payment
This sub-clause makes it clear that any certificate of payment except the final certificate of payment is capable of adjustment and correction and is not conclusive evidence of matters stated therein.

39.9 Application for final certificate of payment
The Contractor can apply for a final certificate of payment once he has complied with all his obligations to make good defects during the Defects Liability Period. If separate Taking-Over Certificates have been issued for Sections of the Works the Contractor can apply for a separate final certificate in respect of that Section once its individual Defects Liability Period has expired. Application for the last final certificate must be accompanied by a final account which includes a detailed analysis and evaluation of all claims that the Contractor may have under the Contract.

39.10 Value of final certificate of payment
This sub-clause makes it clear that all adjustments to the Contract Price must be included in a final certificate of payment. This includes any additions or deductions for variations, or on account of claims by the Contractor or by the Purchaser such as claims for delay under clause 34 (Delay).

39.11 Issue of final certificate of payment
The Engineer must issue a final certificate of payment within 30 days after receiving an application from the Contractor which complies with sub-clause 39.9. It is made clear, however, that this 30-day period will not commence until the Contractor has provided the Engineer with all information amplifying the final account that the Engineer may reasonably require. This does, in fact, place the burden on the Engineer to satisfy himself as to whether he has all the information he requires to enable him to agree the final account and issue the final certificate of payment within 30 days after the Contractor has made his application under sub-clause 39.9.

39.12 Effect of final certificate of payment
The intention of this sub-clause is to try to ensure that, as far as possible, the final certificate of payment really is final and evidences the proper fulfilment by both parties of their obligations under the Contract. However, clearly it would be unfair for the final certificate of payment to be conclusive if fraud or dishonesty affected any matter covered by the certificate. If any proceedings, including arbitration and/or adjudication (if provision for such is made in the Special Conditions) and proceedings in a court of law, have been commenced as at the date the final certificate of payment is issued or if any such proceedings are commenced within three months of the issue of a final certificate of payment, the final certificate of payment is not conclusive.

39.13 No effect in the case of gross misconduct
This sub-clause makes it clear that the Purchaser's rights against the Contractor in relation to latent defects under sub-clause 36.10 (Latent defects) are unaffected by the issue of a final certificate of payment.

40.1   Payment

As previously noted, there are no terms of payment in MF/1 and it is up to the parties to agree on the appropriate terms in the Special Conditions. Nevertheless, this sub-clause requires a Purchaser to pay certificates of payment within 30 days after they are issued. This is a fall back provision which will only apply in the absence of other provisions in the Special Conditions. The second paragraph of the sub-clause makes it clear that if the Special Conditions require the Purchaser to make payments before delivery which are not for work done, then he is entitled to require the Contractor to furnish a bond or guarantee in respect of such payments, against the possibility that the Plant will never be delivered. Such a bond is not required for payment for Plant in the course of manufacture because such payment will only be made under the provisions of clause 37 (Vesting of plant, and contractor's equipment) subject to the transfer of ownership thereof.

40.2   Delayed payment

This sub-clause entitles the Contractor to receive interest on delayed payments. A rate of interest of two percent per annum above the average base rates of the London clearing banks is to apply, unless otherwise agreed in the Special Conditions.

The statutory rate of interest under the Late Payment of Commercial Debts (Interest) Act 1998 is not adopted since it is considered that the right to stop work and/or to terminate the Contract under sub-clause 40.3 would constitute "substantial remedies" for the purposes of that Act.

40.3   Remedies on failure to certify or make payment

This sub-clause permits the Contractor to exercise certain remedies in the event that the Purchaser fails to make payment in accordance with the Contract or the Engineer fails to issue a certificate of payment to which the Contractor is entitled. The Contractor can either, after giving 14 days' notice, stop the Works, in which event he is entitled to recover the Cost of so doing and of any subsequent resumption of work and, under sub-clause 41.2 (Allowance for profit on claims), an element of profit, or he can give notice to the Purchaser to terminate the Contract by 30 days' notice to the Engineer and the Purchaser.

41.   CLAIMS

41.1   Notification of claims

The intention behind the specific procedure for claims for additional payment is that the Purchaser is entitled to be told at the earliest possible opportunity of any circumstances which may give rise to claims for additional payment by the Contractor. It is the intention that unless the procedures specified in this sub-clause are followed, the Purchaser should not be liable to make payment for any claims for additional payment. It will thus be incumbent upon Contractors to introduce the necessary discipline to enable them to comply with the procedures. It should be noted that within 30 days after circumstances arise which the Contractor considers entitle him to claim, he must give notice to the Engineer of his intention to make a claim, with reasons. The Contractor does not have, at that stage, to give full particulars of the amount of the claim and, indeed, in many cases it simply would not be possible for him to do so. All he has to do is to tell the Engineer that he intends to claim and the reasons why. Under sub-paragraph (b) of this sub-clause the Contractor is required as soon as reasonably practicable, and at the very latest not later than the expiry of the last Defects Liability Period, to give the Engineer, with copies for the Purchaser, full particulars of the claim and its amount. Thereafter, he must submit promptly any further particulars that the Engineer may reasonably require to assess the value of the claim. If the Contractor does not comply with the provisions of sub-clause 41.1 he may, by reason of the provisions of sub-clause 41.3, lose his right to payment.

41.2   Allowance for profit on claims

Where the Contractor's claim is based on certain sub-clauses of the Contract the Contractor is entitled to an allowance for profit in addition to the additional Cost he has incurred. This is to be calculated by applying to the total of such Cost the percentage for profit stated in the Appendix. It should be noted that not all grounds for payment of additional Cost will necessarily entitle the Contractor to claim profit. Essentially, the Contractor will only be entitled to profit in those cases where the Purchaser could be said to be in breach, either because of the Purchaser's own action or inaction, or by reason of instructions, lack of instructions or some act or omission of the Engineer.

41.3   **Purchaser's liability to pay claims**

It is intended that the Contractor should lose his right to claims for additional payment unless such claims have been made strictly in accordance with sub-clause 41.1.

42.   **PATENT RIGHTS, ETC.**

42.1, **42.**2, **42.**3 and **42.**4   **Indemnity against infringement, Conduct of proceedings, Purchaser's indemnity against infringement** and **Infringement preventing performance**

These provisions closely follow those included in the previous model forms in the "A" series. It should be noted, however, that under sub-clause 42.3, the indemnity given to the Contractor by the Purchaser against the possibility of designs or instructions from him or from the Engineer causing the Contractor to infringe intellectual property rights, has now been expanded so as to provide the Purchaser with the opportunity of conducting any proceedings, negotiations or litigation in relation thereto in exactly the same manner as the Contractor is entitled to do in the case of infringement of intellectual property rights caused by the Purchaser.

Sub-clause 42.4 provides a remedy for the Purchaser or Contractor, as the case may be, in circumstances where the Contractor is prevented from executing the Works or the Purchaser is prevented from using the Works in consequence of the infringement of intellectual property rights belonging to a third party. Prevention would normally occur by reason of an injunction obtained by a third party against the Contractor or Purchaser as the case may be. In such circumstances under the sub-clause, the party at fault has 90 days from the receipt of written notice to procure, at his own expense, the removal of the cause of prevention. The party required so to do would, in all probability, either attempt to have the injunction removed, if the facts so warranted it, or seek a licence to enable the Purchaser or Contractor, as the case may be, to continue to use the Works or to continue to execute them. If the Purchaser or Contractor is unable so to do, then in the first case the Contractor can treat the Purchaser as being in default and terminate the Contract under clause 51 (Purchaser's default), and in the second the Purchaser may treat the Contractor as being in default and terminate the Contract under clause 49 (Contractor's default).

43.   **ACCIDENTS AND DAMAGE**

The main difference between this clause and corresponding clauses in model forms in the "A" series is the distinct separation from the insurance provisions of the Contractor's responsibility for care of the Works and his liability for loss or damage to the Works and to third party property and for injury to persons and property. MF/1 provides a clear basis of the responsibility and liability which is totally unaffected by any insurance arrangements that may be required in connection with the Works.

43.1   **Care of the works**

The responsibility for care of the Works or any Section lies with the Contractor until a Taking-Over Certificate is issued for the Works or the relevant Section. Equally the Contractor is responsible for the care of outstanding work which he has undertaken to carry out during the Defects Liability Period. If the Contract is terminated in accordance with the Conditions responsibility for the care of the Works passes to the Purchaser when the notice of termination has expired.

43.2   **Making good loss or damage to the works**

If any part of the Works is damaged whilst the Contractor has responsibility for the care of the Works he must make good such loss or damage at his own expense unless it is caused by the Purchaser's Risks [see clause 45 (Purchaser's risks)]. Equally the Contractor must make good at his own expense any loss or damage caused after taking-over whilst completing any outstanding work or in complying with his obligations in relation to defects under clause 36 (Defects liability).

43.3   **Damage to works caused by purchaser's risks**

If any part of the Works is damaged by reason of the Purchaser's Risks while the Contractor has responsibility for care of the Works, the Purchaser has six months to require the Contractor to make good the loss or damage at the Purchaser's expense. If a price cannot be agreed then it is to be determined by arbitration under clause 52 (Disputes and arbitration).

43.4 Injury to persons or damage to property whilst contractor has responsibility for care of the works

The Contractor is liable for and indemnifies the Purchaser against all claims for personal injury or death or for loss or damage to any property (except parts of the Works not yet taken over which are covered by the Contractor's obligations under sub-clauses 43.2 or 43.3 as appropriate) which arises out of or in consequence of the execution of the Works. It should be noted that the Contractor's liability is not dependent upon any negligence on his part and his liability is therefore strict and no proof of fault is required. However, to the extent that death, personal injury or damage results from the Purchaser's Risks, the Purchaser indemnifies the Contractor against any such claims and if damage is the inevitable consequence of the execution of the Works he must similarly indemnify the Contractor.

43.5 Injury to persons or damage to property after responsibility for care of the works passes to purchaser

Essentially this sub-clause governs the Contractor's liability once the Works have been taken over. It does not apply to loss or damage to the Works themselves since the Contractor's liability in respect of those matters is governed by clause 36 (Defects liability). Once the Works have been taken over, therefore, claims for death or injury or damage to any property (except the Works, but including other property of the Purchaser) are the responsibility of the Purchaser but are the subject of an indemnity from the Contractor in favour of the Purchaser to the extent caused by the negligence or breach of statutory duty of the Contractor and those for whom he is responsible or by defective design, materials or workmanship (but not the Purchaser's or Engineer's designs, responsibility for which the Contractor has disclaimed in writing).

43.6 Accidents or injury to workmen

The Contractor is responsible for and must indemnify the Purchaser against any claims made by the Contractor or his Sub-Contractor's employees against the Purchaser for death or personal injury. However, to the extent that the Purchaser may have by his act or default caused that death or personal injury, he must indemnify the Contractor.

43.7 Claims in respect of injury to persons or damage to property

This sub-clause is a procedural provision enabling the Contractor to take over the conduct of all negotiations for the settlement of claims and litigation in relation thereto for which he may be responsible under the Contract. The Purchaser can require the Contractor as a condition of enabling the Contractor to conduct the negotiations, etc., to give security for any costs and damages that may be involved. The Purchaser must, however, give the Contractor all available assistance and is entitled to be repaid all expenses reasonably incurred.

44. LIMITATIONS OF LIABILITY

The provisions of this clause seek to limit the liability of both the Contractor and Purchaser for all matters which lie within their respective responsibilities under the Contract. The philosophy of MF/1 in the area of liability and responsibility is that the Contract between the parties establishes the scope of their individual responsibilities and their liabilities to each other exclusively, with the object and intention that the parties should not be entitled to seek any remedy for a matter upon any basis which is not found in the Contract itself.

44.1 Mitigation of loss

This sub-clause restates the duty of a party to the Contract who has suffered loss to take all reasonable measures to mitigate his loss.

44.2 Indirect or consequential damage

Apart from specific provisions in the Conditions which entitle either the Purchaser or Contractor to claim loss of profit or what might otherwise be commonly termed "consequential loss", this sub-clause makes it clear that neither the Contractor nor the Purchaser will be entitled to claim for loss of profit, loss of use, loss of production, loss of contracts, for financial loss or for any indirect or consequential damage. The reason for this is essentially that, in practice, a party who suffers losses of that nature is in the best position to insure against such losses should he so choose and is in a better position to bear the liability should it arise. For example, there is no reason why the Contractor should be expected to bear all the risks of the Purchaser's decision to invest in the Works or the risks of the operation of the Works when completed, just as there is no reason why the Purchaser should bear all the risks of the Contractor's decision to accept the Contract for the construction of the Works.

44.3   Limitation of contractor's liability

This sub-clause imposes, in relation to any one act or default, an overall limitation of the Contractor's liability to the Purchaser to the sums stated in the Appendix or, if the Appendix has not been completed, to the Contract Price. The parties should consider carefully whether the Contract Price is the appropriate limit for the Contractor's liability for any one act or default and, if necessary, complete the Appendix accordingly. The last sentence of this sub-clause provides that once the Defects Liability Period has expired the Contractor will not be liable to the Purchaser for any loss or damage to the Purchaser's property except to the extent that sub-clause 36.10 (Latent defects) applies. The reason for this is quite simply that the Purchaser is in a better position to insure his property at reasonable premiums against accidental loss or damage and to avoid the necessity of double insurance of such risks.

44.4   Exclusive remedies

The intention of this sub-clause is to make it clear that the parties may only seek their remedies against each other in connection with the Contract or the Works under the Contract itself and are confined to the remedies which the Contract itself provides.

As explained in the introduction to this Commentary, unless a right or remedy is expressly included in the Contract, it cannot be relied upon. If, for example, a decision, instruction or order of the Engineer is to give the Contractor a right to payment of extra Cost, that right to payment must be found expressed in the general conditions or the Special Conditions. Any decisions, instructions or orders may well be "variations" under clause 27 (Variations) but to count as such they must involve an alteration to the Works. The definition of "Works" is such as to extend only to the Plant to be provided under the Contract and the work that the Contractor is required to do under the Contract. It does not, for example, include the facilities to be provided by the Purchaser on Site nor extend to their locations. Any alteration of either would not be a variation and if the Contractor is to recover any extra costs that might be involved, he should ensure that what is to be provided by the Purchaser or what he has assumed will be provided by the Purchaser is expressly included in the Contract documents as an obligation of the Purchaser and that there is an express right to the recovery of extra cost that might be incurred in consequence of alterations. Provision for this can be made in the Special Conditions.

45.   PURCHASER'S RISKS

45.1   Purchaser's risks

This clause provides that the Purchaser will bear the risk of and responsibility for matters which are either within his control, such as the Purchaser's use of the Works, or result from his negligence or in respect of which it is considered that the loss should lie where it falls, eg. damage which is the inevitable result of the construction of the Works. The expression "Purchaser's Risks" also includes risks related to Force Majeure unless they can be insured.

46.   FORCE MAJEURE

46.1   Force majeure

A specific definition of "Force Majeure" is included since under English law the expression has little if any effective meaning unless it is defined. Some of the circumstances mentioned can be insured but this is no reason for excluding them from the definition of Force Majeure or for preventing either of the parties from exercising his rights on the occurrence of Force Majeure under the remaining sub-clauses of this clause.

46.2   Notice of force majeure

If a Force Majeure circumstance prevents or delays either of the parties from fulfilling his obligations then, provided he gives notice of the Force Majeure circumstances and of the obligations which he is delayed or prevented from performing, that party is excused performance or punctual performance of the obligation for so long as the circumstances continue.

46.3   Termination for force majeure

If Force Majeure circumstances continue such that either of the parties is excused performance of his obligations for a continuous period of 120 days then it is open to either party at any time thereafter to terminate the Contract provided that at the time of giving notice such performance or punctual performance is still excused.

46.4   Payment on termination for force majeure

If the Contract is terminated under sub-clause 46.3, to the extent that the Contractor has not already been paid, he is entitled to be paid the Contract Value of the Works as at the date of termination. If he has taken delivery of materials or goods which he cannot legally return then he is also entitled to be paid for those provided he has vested the ownership thereof in the Purchaser. However, the Purchaser need not pay for any such materials or goods until they have been delivered to him or to his order. The Contractor is also entitled to recover expenditure reasonably incurred in the expectation of completing the whole of the Works, eg. the total Cost of setting up his organisation on Site, to the extent not covered by the payment for the Contract Value of the Works. The Contractor is also entitled to the Cost of repatriating his staff and workmen employed at the Site and to the Cost of removal and return of Contractor's Equipment to his premises, or elsewhere at no greater Cost.

47 – 48.   INSURANCE

47.1   Insurance of works

Unless arrangements are made to the contrary, the Contractor is required to insure the Works and Contractor's Equipment in the joint names of himself and the Purchaser for its full replacement value from the date of the Letter of Acceptance until 14 days after the date of issue of a Taking-Over Certificate or 14 days after the date when responsibility for the care of the Works passes to the Purchaser. Essentially, the Contractor is required to effect what is known as "Contractors All Risks" insurance cover against loss or damage however arising other than the Purchaser's Risks. That is the Contractor's minimum obligation in relation to the insurance of the Works. It will be sensible for both Contractor and Purchaser to discuss the terms of the policy with their insurance advisers and, in particular, how the obligation to insure to full replacement value is to be satisfied.

It is customary for policies to be issued showing the original total Contract Price - inclusive of free issue materials, customs dues, freight and insurance (if not already included) - as the initial sum insured. This sum should be reviewed at appropriate intervals and adjusted upwards to reflect the full replacement value at all times as required by the Conditions.

The risks required to be insured do not include the Purchaser's Risks. Some of the risks allocated to the Purchaser can be easily insured under a "Contractors All Risks" insurance policy in particular:

(a)   The Purchaser's or Engineer's design;

(b)   The Purchaser's use of the Works;

(c)   The negligence of the Purchaser, Engineer and/or their servants, agents or other contractors; and

(d)   some of the circumstances which fall within the definition of Force Majeure.

It is suggested that an item for discussion with insurers at the tender stage should be the extent to which the Purchaser's Risks and circumstances which are included within the definition of Force Majeure in the Conditions, can themselves be insured under the joint policy and, if so, the Special Conditions should make clear that such matters are to be included within the cover to be provided.

47.2   Extension of works insurance

This sub-clause requires the Contractor to include within its insurance policy for the Works what is known as "extended maintenance cover".

47.3   Application of insurance monies

If an insured loss arises, the Contractor cannot simply pocket the money and evade responsibility in relation to the replacement and repair of the Plant lost, damaged or destroyed.

47.4   Third party insurance

The Conditions provide for separate third party insurance. Most Contractors will cover their third party liability by a global insurance policy and will not effect a special policy for the purposes of the Contract. If the minimum third party insurance cover specified in the Special Conditions is greater than the Contractor's global policy it should be possible for the Contractor to obtain the required increased cover for the particular Contract and he will include in his prices any increased premium resulting therefrom. The last sentence of the sub-clause is included because many insurers will not permit third party insurance in joint names. Accordingly, the Contractor is required to ensure that there is included in his third party insurance policy a clause to the effect that the Purchaser is entitled to an indemnity from the underwriters where an injured party sues the Purchaser direct instead of proceeding against the Contractor in any case in which the Contractor would have been entitled to recover under his third party insurance had he been sued.

47.5   Insurance against accident, etc. to workmen

This sub-clause requires the Contractor and his Sub-Contractors to have appropriate employers liability insurance which contains an "indemnity to principals clause" in the same manner as is required for the Contractor's third party insurance.

47.6   General insurance requirements

This sub-clause sets out the administrative procedures by which the Contractor satisfies the Engineer and the Purchaser of the existence and maintenance of the insurance cover required. The insurance must be taken out with an insurer and in terms approved by the Purchaser. It should be remembered that in some countries insurance through or by a local insurance company is mandatory. Any alteration to the terms of the policy must be notified by the Contractor to the Purchaser as must any alteration in the amounts for which insurance is provided.

47.7   Exclusions from insurance cover

This sub-clause sets out the exclusions in insurance policies to which the Purchaser is not entitled to object in giving any approval to insurance effected by the Contractor. Paragraph (a) means that in relation to the insurance of the Works the Contractor is not required to effect what is known as "products guarantee" insurance. Such insurance is prohibitively expensive. The remaining permitted exclusions are those risks for which cover is normally excluded by the terms of standard policies. They include risks which insurers are not prepared to insure against at all or without additional and perhaps unrealistic premiums. Motor vehicle risks are excluded upon the basis that such risks will be covered by specific vehicle insurance as required by law.

48.1   Remedy on failure to insure

This sub-clause gives the Purchaser power to effect the insurances himself if the Contractor fails to provide evidence of insurance cover. In such event the premiums paid by the Purchaser are to be deducted from the Contract Price.

48.2   Joint Insurances

This sub-clause is a purely technical provision which the Contractor will need to discuss with his insurance advisors. It is included to ensure that the underwriters of a particular joint insurance policy must waive their rights of subrogation against the other party. In other words, should the Works be damaged before taking-over through the negligence of the Purchaser, the insurers cannot, following payment of the claim to the Contractor, take proceedings against the Purchaser (the other joint insured under the policy) for the purpose of recovering the sums they have paid to the Contractor. The sub-clause also permits the co-insured or the other party to the Contract to be joined as a party in any negotiations, litigation or arbitration relating to the interpretation of the policy or relating to any claim. Since both Contractor and Purchaser have the benefit of the insurance cover both must be entitled to have a say in the settlement of any claims made under the policy or in any dispute as to the meaning of the policy.

49 – 50. CONTRACTOR'S DEFAULT

49.1   Contractor's default

This sub-clause defines those circumstances of serious default on the part of the Contractor which entitle the Purchaser to terminate the Contract and, if he so chooses, to complete the Works himself.

These circumstances include the assignment of the Contract without consent or sub-letting the whole of the Works without consent, rejection of the Works by the Engineer as permitted by the Conditions, abandonment of the Contract by the Contractor, suspension of work entirely without reasonable excuse and failure to proceed for thirty days or more after notice from the Engineer to do so or generally, despite previous warnings, failure to execute the Works in accordance with the Contract, or failing to proceed with due diligence. In any such circumstances the Purchaser can, after having given 21 days' notice to the Contractor, terminate the Contract and expel him from the Site and complete the Works himself. The sub-clause makes it clear that the powers of the Engineer and Purchaser are not affected by such a termination and the Engineer is still entitled to act as the Engineer following termination.

49.2   Valuation at date of termination
If the Purchaser exercises his right to terminate the Contract under sub-clause 49.1 the Engineer is required immediately to value the part of the Works executed and all sums then due to the Contractor. The sum certified as then due is called "the Termination Value".

49.3   Payment after termination
Notwithstanding that the Engineer has certified the Termination Value, the Contractor cannot require the Purchaser to pay that value until the cost of completing the Works has been ascertained and the amount is certified by the Engineer. This is called the "Cost of Completion". The sub-clause then provides for the difference between the Cost of Completion and the Termination Value to be paid to the Purchaser or Contractor as the case may be.

50.1   Bankruptcy and insolvency
This clause permits the Purchaser upon the bankruptcy or insolvency, etc. of the Contractor either to terminate the Contract or to give the receiver, manager, liquidator or administrator the opportunity of taking over and completing the Contract. This may not be possible where the personal ability of the Contractor is relevant to the Purchaser. In other cases bankruptcy or liquidation will usually result in other breaches of contract by delay, etc. so that it may be preferable for the Purchaser to exercise his rights under clause 49 for the reasons therein stated rather than because of the bankruptcy or liquidation itself.

51.   PURCHASER'S DEFAULT

51.1   Notice of termination due to purchaser's default
The Contractor is entitled to terminate the Contract by 14 days' notice to the Purchaser if the Purchaser fails to pay the amount due under a certificate within thirty days after its date of issue. If a longer period for payment of certificates is agreed in the Special Conditions, paragraph (a) of this sub-clause 51.1 should be amended accordingly. The Contractor can also terminate if the Purchaser interferes with or obstructs the issue of any certificate by the Engineer, if the Purchaser becomes bankrupt or goes into liquidation, etc. or if he appoints a person to act with or in replacement of the Engineer against the Contractor's reasonable objections.

51.2   Removal of contractor's equipment
Once the Contractor is given notice under sub-clause 51.1 he is required to remove all Contractor's Equipment from the Site.

51.3   Payment on termination due to purchaser's default
If the Contractor exercises his rights under sub-clause 51.1 to terminate the Contract he is entitled to payment as provided in this sub-clause. To the extent that he has not already been paid he is entitled to receive the Termination Value and to recover expenditure reasonably incurred in the expectation of the performance of the Contract and which has not otherwise been recovered and also expenditure reasonably incurred in consequence of termination. Additionally he is entitled to loss of profit at the rate stated in the Appendix on the difference between the total of those sums and the Contract Price.

52. DISPUTES AND ARBITRATION

52.1 Notice of arbitration

Disputes concerning the Engineer's decisions cannot be referred to arbitration unless the procedures laid down by clause 2 (Engineer and engineer's representative) have been exhausted. Users should note in this connection the specific time limits under sub-clause 2.6 (Disputing engineer's decisions, instructions and orders) because, unless those are observed, the right to challenge the Engineer's decision will be exhausted. Under English law, formal notice of arbitration sent by one party to another constitutes commencement of the arbitration proceedings notwithstanding that the parties may have been unable to agree the arbitrator to be appointed. (The importance of completing the Appendix to the general conditions in this respect, ie. naming the institution which will have responsibility for appointing the arbitrator *before* the need to appoint arises, is emphasised here.) Under English law it is important to ensure that arbitration agreements are in writing. If copies of the general conditions are not referred to in and bound into the contract documents but merely sought to be incorporated by reference, users are advised to ensure that the MF/1 Form of Agreement which specifically refers to the incorporation of clause 52 is used.

52.2 Performance to continue during arbitration

Notwithstanding that arbitration proceedings may have commenced, performance of the Contract is to continue unless the Engineer orders work to be suspended. If work is suspended then the Contractor will be entitled to recover the reasonable Cost incurred and occasioned by the suspension and, under sub-clause 41.2 (Allowance for profit on claims), a profit element. The last sentence of the sub-clause provides that the fact that an arbitration is pending is not to give the Purchaser grounds for withholding any payment which may have become due or is otherwise payable.

52.3 Arbitrator's powers

Specific power needs to be granted to the arbitrator to enable him to substitute his own views on matters which would otherwise solely be within the Engineer's discretion such as certificates or valuations. Specific power is also granted to the arbitrator to make interim awards. This specific power makes clear that the granting of interim relief is not restricted to the courts.

For Scots law contracts the amendment to this sub-clause suggested in the Special Conditions gives the arbiter the necessary express power to assess damages which he would not otherwise have.

52.4 Joinder

This clause is the counterpart to sub-clause 19.3 of the Form of Sub-Contract. It is clearly desirable that if, in essence, a dispute between the Purchaser and the Contractor is in whole or in part dependent on the resolution of a related dispute between the Contractor and the Sub-Contractor that such matters should be determined at the same time in the same proceedings.

For Scots law contracts the amendments to this sub-clause suggested in the Special Conditions are purely a matter of terminology.

52.5 Arbitration rules

MF/1 does not lay down any particular rules for the conduct of the arbitration such as those published by the International Chamber of Commerce (ICC) or the Chartered Institute of Arbitrators (CIArb) and, accordingly, unless there is some provision in the Special Conditions requiring arbitration to be conducted under the specific arbitration rules of a particular body, any arbitration will need to be conducted in accordance with the Arbitration Act 1996. Where MF/1 is used for overseas contracts it may be appropriate to adopt either the Rules of Conciliation and Arbitration of the ICC or the UNCITRAL Arbitration Rules. In such circumstances, appropriate advice on the possible amendment of sub-clause 52.1 should be taken. For contracts under English law the Construction Industry Model Arbitration Rules (CIMAR) may be appropriate.

52.6 Consent to registration

(See the suggestions for provisions relating to "Disputes and arbitration" in the Special Conditions under the sub-heading of "Scots law".) This sub-clause makes the arbiter's decision more readily enforceable without recourse to the courts since, in Scotland, an arbiter has no jurisdiction to enforce his award.

53. SUB-CONTRACTORS, ETC.

53.1 Sub-contractors, servants and agents

The purpose of this clause is to exclude any right of the Purchaser to take action direct against servants or agents of the Contractor or against any Sub-Contractor for any loss, damage or injury however this may have arisen. The purpose of this provision is to make it abundantly clear that the Purchaser's remedies lie against the Contractor and no one else. The clause also specifically states that all limitations and exclusions of liability provided for the Contractor under the Conditions are also extended to protect servants or agents of the Contractor and his Sub-Contractors.

54. APPLICABLE LAW

54.1 Applicable law

The Contract and the Conditions have all been prepared on the basis that they are governed by English law and that the procedural law for any arbitration will also be English law. Users wishing to apply Scots law should note the amendments required in such circumstances to clause 52 (Disputes and arbitration) given in the Special Conditions. Users who wish to apply laws other than English law or Scots law should obtain appropriate legal advice.

4. Appendix to the general conditions

The details which are required to be inserted into the Appendix to the General Conditions must be completed by the Purchaser prior to inviting tenders or be negotiated for each Contract. In any event the Appendix must be completed and agreed before a Letter of Acceptance is issued.

5. Special conditions

Aide-mémoire to their preparation

The details printed under this heading in MF/1 are solely an aide-mémoire to the preparation of the Special Conditions for particular contracts. It will be noted that there are included under the heading of Special Conditions some suggested clauses providing for progress certificates of payment and suggestions for terms of payment. The headings of the various items under the Special Conditions are by no means exhaustive and in most cases the Engineer will need to consult with or advise the Purchaser and/or vice versa on the content of the Special Conditions.

ADDITIONAL SPECIAL CONDITIONS FOR USE IN CONTRACTS INVOLVING THE INCIDENTAL SUPPLY OF HARDWARE AND SOFTWARE

Detailed suggestions are included under this heading in MF/1 to provide users with detailed guidelines or suggestions for the Special Conditions which may be appropriate where the Works involve the provision of computer hardware and software. It is emphasised that these suggested additional Special Conditions are a model only.

ADDITIONAL SPECIAL CONDITIONS FOR USE IN CONTRACTS WHERE CERTIFICATION FOR PAYMENT AND PAYMENTS ARE TO BE DETERMINED IN FULL OR IN PART BY MEASUREMENT

Special Conditions are needed for use in contracts where the principal method by which the Contractor's entitlement to payment is determined by measurement. The Special Conditions suggested include the revision of several sub-clauses in MF/1 and the provision of additional clauses which are considered appropriate.

It is emphasised that the suggested additional Special Conditions are a model only and should be adapted and supplemented where necessary to reflect the nature of the Works and the problems which are likely to be encountered. In particular, it may be necessary to expand the definition of the Site, where the Works involve pipelines, powerlines or cabling contracts, to reflect the fact that such work often takes place over a lengthy route.

The measured work form of contract is particularly appropriate for pipeline, powerline, cable, and similar contracts where unforeseen difficulties such as hidden obstacles are commonplace, the rights of land owners must be respected (and their goodwill retained) and additional statutory or other regulations may apply. The Works will, in general, be subject to planning approvals which go beyond those which would apply to a Site at a single location and the Works may need to comply with statutory regulations in respect of highways, railways and safety of the public, etc. The Contractor will frequently find that minor changes to the route are necessary and will encounter more or less difficult ground conditions. Such changes can all be addressed effectively by payment against measured quantities.

The Special Conditions suggest changes to sub-clause 5.2 (Site data) and sub-clause 5.7 (Unexpected site conditions) which acknowledge that the Contractor may expect to meet buried obstructions such as sewers, drains, etc. which he may be instructed to divert or remove as a variation to the Contract and that unexpected physical conditions on an extended Site will not be restricted to those below ground.

Additional provisions are included in sub-clause 6.1 (Statutory and other regulations), sub-clause 11.2 (Wayleaves, consents, etc.), new sub-clause 11.9 (Third party interests) and new clause 7 (Statutory and other requirements) to reflect the nature of the extended Site with regard to the duties owed to grantors of wayleaves, land owners on or adjacent to the route and others who may be affected by the Works. A new clause 9 (Reinstatement) covers making good unavoidable damage during construction which is particularly relevant to work carried out in or adjacent to highways.

It may also be necessary to require the Contractor to record accurately the route and its depth or height in relation to a given datum. Whilst such requirements will normally be included in the Specification, it may be appropriate to deal with them additionally in the Special Conditions.

It should be appreciated that where the Specification requires the Contractor to obtain drawings, information and records from third parties, what can be made available may not be complete or may be inaccurate. Obvious errors or omissions should be referred to the Engineer by the Contractor for instructions. Where, however, the Contractor would be expected to accept in good faith and to interpolate such information as he has, then providing he adopts a reasonable interpolation the Contractor should not be responsible if the interpolation proves not to be correct.

Users are advised to take care when defining the units of work for which payment will be made and the method of measurement. Particular care should be taken to specify the full extent of each unit of work and to define clearly the boundaries between the work units. Any general principles adopted in separating one unit of work from another should be clearly stated. Similarly, it is advisable to define the general principles to be adopted when measuring work as well as the actual method of measurement.

ADDITIONAL SPECIAL CONDITIONS FOR USE WHERE THE CONTRACT IS TO PROVIDE SECTIONAL COMPLETION AND DAMAGES FOR DELAY IN COMPLETION OF SECTIONS

Detailed suggestions are included under this heading to provide users with suggested amendments to clauses 32.1 (Time for completion) and 34 (Delay) for use in circumstances where damages are payable for failure to complete a Section for which a specific Time for Completion has been fixed in the Contract. Users are reminded that it is normally inappropriate to agree liquidated damages for failure to achieve sectional completion unless the Section, when completed, is capable of beneficial use.

ADDITIONAL SPECIAL CONDITIONS FOR USE IN CONTRACTS WHICH ARE SUBJECT TO THE HOUSING GRANTS, CONSTRUCTION AND REGENERATION ACT 1996 ("the Act")

Where a contract let on MF/1 Conditions is to be performed in the United Kingdom, it is likely that such a contract, being one for carrying out "construction operations" as defined in the Act, will be liable to the provisions of the Act. Certain types of contract normally let on MF/1 Conditions are excluded from the Act's operation, in particular contracts which are to be performed on a Site where the primary activity is nuclear processing, power generation or water or effluent treatment or the production, transmission, processing or bulk storage (but not warehousing) of chemicals, pharmaceuticals, oil, gas, steel or food and drink. The applicability of the Act is far from straightforward and legal advice may be needed to establish whether a particular contract is subject to the Act.

The Act, amongst other things, gives the parties to "construction contracts" the right to have their disputes adjudicated at any time (in addition to any provisions in the contract for disputes to be determined by the courts or in arbitration), prohibits conditional payment provisions (eg. "pay when paid" clauses) and makes certain other provisions in relation to payment. If the relevant provisions required by the Act are not included in a construction contract then a statutory scheme implies the appropriate provision into the contract (eg. under the Scheme for Construction Contracts or the Scheme for Construction Contracts (Scotland)).

The provisions of the Act are extended to Northern Ireland by the Construction Contracts (Northern Ireland) Order 1997 (the Order). Where the Contract is to be performed in Northern Ireland and is treated under the Order as a construction contract the Order and equivalent Articles should be substituted for the Act and for the referenced sections of the Act.

As MF/1 is a model form of contract designed for international use and the Act only applies to contracts to be performed in the United Kingdom, the Act's requirements are addressed by the additional Special Conditions.

ADDITIONAL SPECIAL CONDITIONS FOR USE IN CONTRACTS WHICH ARE SUBJECT TO THE CONTRACTS (RIGHTS OF THIRD PARTIES) ACT 1999

As a consequence of this Act, contracting parties can confer a benefit on a third party. Parties may wish to:-

(i) ensure from the outset that a third party cannot acquire unintended rights under the Contract. An appropriate provision should be included in the Contract and, as required, a similar provision should be included in any MF/1 Sub-Contract(s). Suitable suggested Special Conditions are provided.

(ii) confer a benefit of the Contract on a third party. Again, parties should make an appropriate provision in the Contract by means of a Special Condition and the terms of any Sub-Contract(s) should not be overlooked in relation to the Act.

6. The Form of Sub-Contract

MF/1 includes a Form of Sub-Contract suitable for use when the Main Contract is under MF/1 Conditions. The Form of Sub-Contract has been designed to dovetail in with MF/1 Conditions and so to provide the Contractor with a Form of Sub-Contract which is "back to back" with his Main Contract. A similar drafting convention to that used in the general conditions of the Main Contract, see comment on clause 1 (Definitions and interpretations) in section 3 of this Commentary, is adopted in the Form of Sub-Contract but with a differing set of defined key terms.

It is emphasised that the MF/1 Form of Sub-Contract, when completed, is a binding contract collateral to the Main Contract and so users should exercise appropriate care in deciding its details.

The Sub-Contract has been prepared on the basis that it is governed by English law and that the procedural law for any arbitration will also be English law. However, it may be adapted for use under other laws by the inclusion of appropriate provisions within the Ninth Schedule. Users wishing to apply Scots law should note the amendments required in such circumstances to sub-clause 52.1 (Notice of arbitration) given in the Special Conditions of the Main Contract. If the parties wish to substitute another substantive law, other than Scots law, either to govern the Sub-Contract itself or to govern procedures under the arbitration agreement, they should seek appropriate legal advice.

Particular points which should be borne in mind by both Contractor and Sub-Contractor in determining whether a Sub-Contract should be entered into using this form are as follows:

(a) Where the Main Contract provides for liability to be limited by reference to the Contract Price the Sub-Contractor's liability under sub-clause 3.3 will be limited by reference to the Sub-Contract Price. Thus if the Sub-Contractor is totally responsible for a liability so limited under the Main Contract the amount which the Contractor is entitled to recover from the Sub-Contractor may be insufficient to cover the Contractor's liability to the Purchaser.

(b) The Sub-Contractor's total liability under the Sub-Contract under sub-clause 3.4 is not to exceed the Sub-Contract Price.

(c) The Sub-Contractor's liability for delay under clause 7.1 (Delay in completion) is limited by reference to the appropriate percentage of the Sub-Contract Value of the part of the Works that cannot be used by reason of his delay and his maximum liability is limited by reference to the maximum percentage of the Sub-Contract Price. The Contractor may not, therefore, recover sufficient from the Sub-Contractor to meet the Contractor's liability for delay to the Purchaser under the Main Contract.

(d) The Sub-Contractor is to be included under clause 12 (Insurances) as a co-insured under the insurances required by sub-clauses 47.1 (Insurance of works) and 47.2 (Extension of works insurance) of the Main Contract.

(e) The Sub-Contractor's obligations in relation to defects under clause 13.1 (Defects liability) do not expire until the end of the Defects Liability Period under the Main Contract.

7. Additional documents

Included within MF/1 are a Form of Tender, a Form of Agreement, a Form of Performance Bond, a Form of Defects Liability Demand Guarantee, a Form of Notice of Delegation of Authority, a Form of Variation Order and a Form of Taking-Over Certificate.

It is not considered that the Form of Tender, the Form of Agreement, the Form of Variation Order, the Form of Notice of Delegation of Authority or the Form of Taking-Over Certificate warrant particular comment other than they are suitable for use with the MF/1 Contract.

The Form of Performance Bond and the Form of Defects Liability Demand Guarantee are commented on in paragraphs (i) & (ii) below. It is to be noted that because these two documents use different "keywords" –some of which are taken from their parent publications (in the interests of succinctness of description)- the drafting convention defined in section 3.1 of this Commentary could not be strictly observed.

Both these model forms incorporate International Chamber of Commerce Uniform Rules by reference. The parties should familiarise themselves with the relevant ICC Uniform Rules before utilising these model forms. Copies of the ICC Uniform Rules are available from ICC United Kingdom.

(i) Form of Performance Bond

The model Form of Performance Bond is a conditional bond and is not payable on the first simple demand of the Purchaser. The model form is based on that published by the Association of British Insurers but incorporates by reference the Uniform Rules for Contract Bonds published by the International Chamber of Commerce (ICC) (Publication No. 524). Those rules provide a clear and concise scheme to regulate the nature of the obligations arising under the Bond and the claims procedure and are vital to a proper understanding of the use of the Bond. The Purchaser's right to claim under the Bond arises in the event of default by the Contractor which is specifically stated to include the termination of the Contract by the Purchaser as a consequence of the Contractor's insolvency - see clause 50.1(a). The Purchaser can, however, only claim once the damages he has suffered in consequence have been established and ascertained (by agreement or after court or arbitration proceedings) or determined under sub-clause 49.3 (Payment after termination). The Purchaser must when claiming give credit for all sums which are or may become due to the Contractor.

It may take some time to establish the Purchaser's loss and, accordingly, the Bond will not expire until three months after the date of issue of the final certificate or until three months after any proceedings (in arbitration or through the courts) have been finally settled. The parties are free to agree a different Expiry Date but if they do so they may find that the Bond expires and becomes unenforceable before the loss suffered by the Purchaser can be ascertained.

The ICC rules permit the Guarantor in the event of default to perform the Contractor's obligations under the Contract. It is not thought such an entitlement is one which will be attractive to Purchasers except those used to the operation of the system of surety bonds common in the United States of America. Clause 1 of the Performance Bond accordingly excludes this right of the Guarantor.

(ii) Form of Defects Liability Demand Guarantee

A model Defects Liability Demand Guarantee for use with the suggested terms of payment set out in the aide-mémoire to the preparation of the Special Conditions and applicable to contracts let on MF/1 Conditions has been provided. The Defects Liability Demand Guarantee is payable on the demand of the Purchaser since the sole purpose of the Guarantee is to secure the performance by the Contractor of his obligations in relation to defects and damage caused by defects in circumstances where the Purchaser has agreed to release retentions prior to the issue of the final certificate of payment. The Defects Liability Demand Guarantee incorporates by reference the Uniform Rules for Demand Guarantees published by the International Chamber of Commerce (ICC) (Publication No. 458). Those rules provide a clear and concise scheme to regulate the nature of the obligations arising under the Guarantee and the claims procedure is vital to a proper understanding of the use of the Guarantee. Unlike with the Performance Bond the Purchaser is not required to prove his loss. He must, however, provide an authenticated written demand and a certificate of the circumstances which have given rise to the right to claim. The expiry of demand guarantees is always an important matter and it should be noted that the Guarantee expires when the Guarantor has been notified of the issue of the final certificate of payment under the Contract. The Purchaser is required to return the Guarantee to the Guarantor on expiry. If the parties so wish, an earlier expiry can be chosen, eg. a specific calendar date under the provisions of Article 22 of the ICC Uniform Rules.

8. Useful addresses / sources of information

Publications of the bodies mentioned in either MF/1 or in the Commentary and/or further information can be obtained from the following sources.

1.	Chartered Institute of Arbitrators International Arbitration Centre 24 Angel Gate, City Road LONDON United Kingdom EC1V 2RS	Tel: +44 (0)20-7837 4483 Fax: +44 (0)20-7837 4185 E-mail: info@arbitrators.org	Note 1
2.	Construction Industry Model Arbitration Rules (CIMAR) c/o the Honorary Secretary The Society of Construction Arbitrators Forty One, Rowsham Dell Giffard Park, MILTON KEYNES United Kingdom MK14 5JS	Tel: +44 (0)1908 618845 Fax: +44 (0)1908 216594 E-mail: christopher@ dancaster.demon.co.uk	Note 2
3.	ICC United Kingdom International Chamber of Commerce 14-15 Belgrave Square LONDON United Kingdom SW1X 8PS	Tel: +44 (0)20-7823 2811 Fax: +44 (0)20-7235 5447 E-mail: reneecohen@iccorg.co.uk	Note 3
4.	Joint IMechE/IEE Model Forms Committee Secretary Knowledge Services Engineering Policy Department IEE, Savoy Place LONDON United Kingdom WC2R 0BL	Tel: +44 (0)20-7344 5411 Fax: +44 (0)20-7344 8408 E-mail: modelforms@iee.org.uk	Note 4
5.	Law Society of Scotland 26 Drumsheugh Gardens EDINBURGH Midlothian United Kingdom EH3 7RN	Tel: +44 (0)131 226 7411 Fax: +44 (0)131 225 2934 E-mail: lawscot@lawscot.org.uk	Note 1
6.	London Court of International Arbitration 8 Breams Building LONDON United Kingdom EC4A 1HP	Tel: +44 (0)20-7405 8008 Fax: +44 (0)20-7405 8009 E-mail: lcia@lcia-arbitration.com	Note 5

7.	Scottish Council for International Arbitration c/o MacRoberts, 27 Melville Street EDINBURGH United Kingdom EH3 7JF	Tel: +44 (0)131 220 4776 Fax: +44 (0)131 226 2501 E-mail: maildesk@macroberts.co.uk	Note 6
8.	The Stationery Office PO Box 29 NORWICH United Kingdom NR3 1GN	Tel: +44 (0)870-600 5522, (option 4) Fax: +44 (0)870-600 5533 E-mail: book.orders@theso.co.uk	Note 7
9.	UNCITRAL (United Nations Commission on International Trade Law) Vienna International Centre PO Box 500 A–1400 VIENNA Austria	Tel: +43 1 26060 4060/61 Fax: +43 1 26060 5813 E-mail: uncitral@uncitral.org	Note 8

NOTES

1   For the purchase of copies of their Model Arbitration Rules.

2   Published by the Society of Construction Arbitrators and particularly suited for use with MF/1 on contracts subject to English law.

3   ICC publications are available for purchase from ICC (UK).

4   For general enquiries on the "MF" Series of Model Forms and their associated Commentaries (but not for sales or for "interpretations" of their provisions).

5   The LCIA can provide copies of a range of Model Arbitration Rules and also full facilities for hearings.

6   The SCIA can provide copies of the ICC Arbitration Rules (Edinburgh) and the UNCITRAL Arbitration Rules (Scotland).

7   The Stationery Office publish UK statutes and related material.

8   UNCITRAL can provide copies of their arbitration documents, but do not give practical advice.